D0872996

Content and Comportment

Content and Comportment

On Embodiment and the Epistemic Availability of the World

Michael O'Donovan-Anderson

ROWMAN & LITTLEFIELD PUBLISHERS, INC.
Lanham • Boulder • New York • London

ROWMAN & LITTLEFIELD PUBLISHERS, INC.

Published in the United States of America
by Rowman & Littlefield Publishers, Inc.
4720 Boston Way, Lanham, Maryland 20706

3 Henrietta Street
London WC2E 8LU, England

Copyright © 1997 by Rowman & Littlefield Publishers, Inc.

All rights reserved. No part of this publication may be reproduced,
stored in a retrieval system, or transmitted in any form or by any
means, electronic, mechanical, photocopying, recording, or otherwise,
without the prior permission of the publisher.

British Cataloging in Publication Information Available

Library of Congress Cataloging-in-Publication Data

O'Donovan-Anderson, Michael.
    Content and comportment : on embodiment and the epistemic availability of the
world / Michael O'Donovan-Anderson.
        p.   cm.
    Includes bibliographical references and index.
    ISBN 0-8476-8624-8 (cloth : alk. paper). — ISBN 0-8476-8625-6 (paper : alk.
paper)
    1. Knowledge, Theory of. 2. Body, Human (Philosophy) 3. Intentionality
(Philosophy) 4. Mind and body. I. Title.
    BD201.O36   1997
    121—dc21   97-25294

ISBN 0-8476-8624-8 (cloth : alk. paper)
ISBN 0-8476-8625-6 (pbk. : alk. paper)

Printed in the United States of America

♾™ The paper used in this publication meets the minimum requirements of American
National Standard for Information Sciences—Permanence of Paper for Printed Library
Materials, ANSI Z39.48–1984.

Contents

Preface

It has become a commonplace in philosophy (and in the theoretical humanities more generally) to assert the ultimate inaccessibility of material reality, to interpret the experienced limitations of the human intellect and epistemic capacity as evidence of our cognitive inadequacy to the world. This conviction takes myriad forms—historical, cognitive, social, linguistic—but can be fairly recognized (if not defined) by its metaphors: We are "trapped" by our senses, our historical moment, our race, "limited" by our conceptual schema, our language, our moral conscience. We cannot get "outside" of the web of concepts, "around" the veil of sense, "beyond" the horizon of language to see the world as it is. These metaphors give expression to the notion that our mind "inside" cannot grasp the reality "outside" because we are confined by personal, social, historical, ethical or linguistic walls which, however flexibly they may be bent to the shape of the world, still intrude their impervious bulk between us and the reality we seek to know.

Although this study has more particular motivations and, of necessity, a specific and somewhat limited scope, it is most generally intended as a step in overcoming such metaphors of cognitive confinement, and of the dualism which both supports and is implied by them. It is far from my intention to deny the difficulty and uncertainty of matters of empirical investigation; but I want to insist that we are more open to the world—more epistemically porous—than the picture of our confinement allows. William James, himself acutely aware of the trials and tribulations of Natural Science, nevertheless observes that "[e]xperience . . . has a way of boiling over, and making us correct our present formulations." This seems to me just right, but understanding how it can be so will, I think, require us to put aside our dualistic prejudices, our easy confidence that it is in fact possible (or even necessary!) to understand mind as over against body, thought over against action, or a cognitive "inside" to be opposed to a material "outside" with precise delimitations and impenetrable boundaries.

As I have indicated, I will not be addressing these issues as such (would that

I could!). It must suffice for me here to question the accuracy of the cognitive psychology which matches so well the dualistic picture of epistemic confinement hanging in so many philosophic homes. What I mean to resist is the move to identify our empirical porosity, our epistemic openness, entirely with sensation. I will not deny the plausibility of the vaguely neo-Kantian notion that our senses deliver *something* (impulses? information?) which is arranged, synthesized or otherwise interpreted by large-scale cognitive structures to produce (conceptually structured) beliefs. This picture is probably right in outline, but here, as always, the devil is in the details. When it is supposed that the deliverances of our sense organs constitute the whole of our epistemic contact with the world, then we can retain this plausible account only by conceding our cognitive confinement. This concession is unwise, and, if I am right, also unnecessary, for, important though our sense organs are, they do not provide our only means of knowing the world. I will be arguing for a cognitive and epistemic psychology which gives pride of place to bodily activity, to the behavior of the mindful body. It is in this mindful, embodied activity that I will locate a kind of epistemic openness—another epistemic conduit—which lets the world seep through conceptual boundaries real and imagined, and reveals us as beings cognitively in touch with (because physically *in*) the world.

Cognitive content, this is to say, has much to do with thoughtful comportment. Indeed, I do not think it too bold to say that thinking—at the very least, empirical thinking about physical objects—is something that only embodied beings can do. In developing this insight about the connection between embodiment and thought I hope to find the resources to deny our confinement, and to clear a space for the possibility of objectivity. Although we may never see the world whole, it is possible, perhaps, to see some of its parts truly.

Acknowledgments

In addition to the people noted in the text for their particular contributions to this book, I should like also to thank Karsten Harries, without whom this would be a very different work, and Anthony Appiah, because of whom this is a better work. For helping make this a book at all, thanks are due to Robin Adler and Dorothy Bradley, my editors and Rowman and Littlefield, and to Julia Legas, who prepared the index to this volume.

Over the past six years I have benefited from the company and wisdom of remarkable friends and colleagues, including Sarah Broadie, Richard Capobianco, Maren Holste, Steve Horst, Michele Janette, Ron Katwan, Trip McCrossin, Ernan McMullin, David Ober, Elen Roklina, Timothy Rosenkoetter, Carol Rovane, Susan Shell, and Richard Velkley. Extremely important to me have been the companionship of, and the innumerable hours of discussion with, Colin Sample (whose taste in wine approaches his facility with philosophy). Finally, more gratitude than I can express belongs to Maeve O'Donovan-Anderson, with whom I learned all I know about the intersubjective practice of Reason.

1
Intentionality and the Fourth Dogma of Empiricism

On the one hand, it is supposed, modestly, that how matters stand in the world, what opinions about it are true, is settled independently of whatever germane beliefs are held by actual people. On the other, we presume to think that we are capable of arriving at the right concepts with which to capture at least a substantial part of the truth, and that our cognitive capacities can and do very often put us in a position to know the truth, or at least believe it with ample justification. The unique attraction of realism is the nice balance of feasibility and dignity that it offers to our quest for knowledge. Greater modesty would mean doubts about the capacity of our cognitive procedures to determine what is true—or even about our capacity to conceptualize the truth—and, so, would be a slide in the direction of skepticism. Greater presumption would mean calling into question, one way or another, the autonomy of truth and, so, would be a slide in the direction of idealism. To the extent that we are serious about the pursuit of truth, we are unlikely to be attracted by either of these tendencies. We want the mountain to be climbable, but we also want it to be a real mountain, not some sort of reification of aspects of ourselves.

—Crispin Wright[1]

1.1: Introduction

I once remarked to a colleague that academic work seemed always to take the form either of a shell or of a corner—either a geodesic frame limiting a yet to be addressed center, or the detailed analysis of some small section of what promises to be a much larger canvas. He thought for a moment before replying: "I think I'm working on the corner of a shell."

The work that follows is undoubtedly a shell—and it has all the advantages and disadvantages of that form; I have tried to keep it simple and clear. Not unlike Fuller's architecture it is meant to be primarily suggestive, but also highly structured and (I dare hope) flexible and strong. Because it eschews detail, the work is not site specific, and I believe that analytic epistemology is not the only subdiscipline of the theoretical humanities which could benefit from the analysis offered here.

It is important to realize the extent to which each of the features listed above is *both* an advantage and a disadvantage. A work which aims primarily at clarity is, if successful, a work easy to criticize. I have not tried to avoid this vulnerability by continually referring to, and attempting to defuse, potential counterarguments. In this regard I take as my stylistic model Descartes' *Meditations*. Not that I appear as a character in my narrative (and certainly not in my dressing gown!), but *Meditations on First Philosophy* is among the clearest—and most criticized—works in philosophy. It is also among the most successful. I attribute its influence to an intelligent simplicity and single-mindedness, features I have tried to emulate here. For a complicated analysis deeply involved in the details of some particular subfield is thereby less suggestive, if only because the reality of disciplinary specialization prevents any one for whom an analysis might be suggestive from understanding the suggestions; and even the rare individuals who are multilingual often encounter great difficulties (if not actual incommensurabilities) in translation. Likewise, a detailed paper purporting to be definitive in its realm is for that reason a work aimed at closure. As important as such closure is to the progress (the history?) of any discipline, openings are also crucial. I intend this to be an opening, to present a direction for fruitful thought, and I take the measure of success for such a project to be the fertility of the direction, more than the fate of particular results herein obtained.

Still, there are such results. This is a work on epistemology, albeit one motivated by a problem in semantics. I want to ask, with Wittgenstein: "What makes my thought about him a thought about *him?*" Because it is my (widely shared) intuition that an answer to that question must involve an account of our *knowledge* of him (and I will provide some reasons for preserving this insight) I will be asking, in the service of Wittgenstein's query, the more general: "What is the nature of our epistemic connection to the world which allows for this intentional—referential—connection to material particulars?" What, in the structure of our epistemic faculties accounts for the fact that when we know, we know *about* things in the world? It should be duly noted that I will not be asking the skeptical question: "How do we know that the objects about which we have beliefs and/or knowledge are 'really in' the world?"—for I will be assuming the existence of an actual, articulated world, and investigating instead how and to what degree we have epistemic access to that world.[2] But although I will try to eschew skeptical questions (that mentioned above, and the closely related questions regarding the possibility of systematic error) insofar as this is possible for any epistemological work (and it is not, entirely), it is nonetheless true that my concern with the "aboutness" of our knowledge includes concern for the appropriateness of that knowledge to the world. In part, this is motivated by the obvious insight that inappropriate knowledge could hardly ground intentional or referential connection in any systematic way. Primarily, however, my concern with appropriateness has its independent motivation in my desire to provide a grounding for a realist notion of truth. As is nicely suggested by Crispin Wright in the quote which opens this chapter, we should view the requirements of truth as constraints on epistemic theorizing: all things considered, it would be preferable if a theory of knowledge could simultaneously account for the epistemic availability *of* the world, and for our epistemic openness *to* the world, that is, provide both a convincing model of the form and nature of our epistemic access to the world, and an account of the ways in which the world provides epistemic friction sufficient to limit and direct our conceptions of it. As Wright notes, greater modesty about our epistemic powers implies the real world is unavailable to us, whereas greater presumption might imply conceptualizations which override the world, projecting onto it reifications of our own thought (or, taking the metaphysical implications of this view seriously, the existence of a world dependent on mind-created categories). Whatever the pressures to adopt the skeptical attitude, and whatever the attractions of idealist metaphysics, it is my express intention to avoid both.

It is because the primary aim of this project is to outline an epistemic theory which can meet these metaphysical and semantic constraints that the argument I present has the generality characteristic of a shell. But for this same reason, detailing these constraints becomes central to the project, and the argument must therefore be situated in a tradition which runs from Locke, through the American pragmatists, to such thinkers as Davidson, Putnam and Rorty, utilizing the resources of analytic philosophy of language and continental phenomenology. This book, that is to say, is self-consciously the inheritor of three different and sometimes rival traditions, and it will not be possible to understand fully its perspective without understanding the specific character of its motivations, without at least noticing the idiom in which its concerns are cast. For these traditions, in both providing the tools for, and being the subject of, this inquiry also suggest the desiderata in terms of which the success of this project will ultimately be judged. Thus I will spend the remainder of this introduction making my perspective and motivations apparent.

There is one category of philosophical work on which I spend notably little time, and that is contemporary epistemology itself. This may be a bit surprising, but the lacunae is easily explained: I do take theories in which our epistemic access to the world is limited to the deliverances of the sense organs, and our knowledge is the product of some form and degree of conceptual synthesis of this sensual information, to be overly influential in (analytic) philosophy of knowledge;

however, my quarrel is not with the details of any particular epistemology or set of epistemologies, but with these very general assumptions.[3] I hope to show, first through a detailed look at Frege's semantics, and next through a critical analysis of American pragmatism,[4] that these assumptions force knowledge to be so empirically distanced from the world as to lose the epistemic friction which characterizes empirical content; this threatens the notion that content is determined by, and therefore our knowledge is about, the actual world.[5] To be placed in such an epistemic position is always to be forced to choose between skepticism and idealism, for together the assumptions tend to encourage the conflation of sensual criteria of identification for physical objects and classes with their criteria of identity (because forced to conclude that mind, in virtue of its synthesis of sensual information, "defines" the division of reality). Yet to deny the conflation seems merely to assert the epistemic inaccessibility of an object's criteria of identity, opening the door to skepticism.

I argue that this artificial and unfortunate limitation of metaphysical options can be avoided if we posit a nonsensual mode of openness to the material structure of the world, a mode of epistemic access rooted in the active body.[6] It is something of a consensus opinion that epistemological realism is possible only if the world *does* provide a degree of epistemic friction sufficient to limit and direct the contents of our knowledge; much epistemology seems committed to finding this friction in sensation. I argue that this is a mistake, that sensation is the wrong place to look for the epistemic friction the world provides. Instead, I argue that the world's influence on the contents of our knowledge is to be found in activity, in the limitation and direction of our behavior by the structure of the physical world. In particular, I maintain that although concepts are called into operation in empirical thinking by sensory contact with the world, they are shaped by, and to some degree defined in terms of, our *comportmental* encounter with the world.

But although the epistemic issues cited above motivate this project, it is in no way the task of this work to refute either skepticism or idealism.[7] The book is not even *about* idealism, skepticism, or even realism, if only because of the problem of scope this would present: the myriad realisms and idealisms in philosophy make immediate mockery of any claim to address them adequately.[8] The task of this work is to inquire about the nature and scope of our epistemic access (and openness) to the world, and to develop what I take to be a neglected mode of such access. None of this is to say that this inquiry makes no contribution to the case for realism; I *do* claim that the mode of epistemic access to the world which I uncover is sufficient to ground the claim that we have knowledge *of* the world, and that we are capable of judging the appropriateness of our knowledge *to* the world. That is, I do insist that the mode of access I uncover is, indeed, access. Justifying this claim—and establishing what it takes to justify it—is the main burden of this book.

1.2: Knowledge and Epistemic Access

We (humans) possess a great deal of knowledge; we have beliefs, impressions, theories, and the like—a great majority of which are, it seems, true. This is a datum in need of some explanation. In virtue of what can we account for our capacity to produce knowledge reliably, knowledge which allows us to negotiate the world, build bridges, locate, identify and describe objects, communicate with one another, and otherwise understand our world?

We might note immediately that it seems a criterion of (being) knowledge to have an object—knowledge is always about things, about, vaguely speaking, the world. One needn't be in the business of reducing truth to more fundamental notions (abstract relations like correspondence, or practical criteria like utility) to see that truth must therefore be understood as an assessment of the appropriateness of that relation of "aboutness." It is one explicit aim of this work to present an epistemic frame within which this simple intuition about truth can be preserved. This aim is, of course, hardly novel. Indeed, it would be very difficult to make sense of the history of epistemology without at least seeing the attraction of this insight. For if we take the most likely explanation of our successes in science, of the ease with which we negotiate our world, to be that most of what we believe is, indeed, true (and we have, historically, taken this to be so), then only if we take the touchstone of truth to be its acknowledgment of the appropriateness of our knowledge to the world will we further take our reliable production of knowledge to indicate a capacity to access that world. We know about the world because we have access to the world. Thus has it been the job of epistemology to account for, to explain, that (apparent) access.

I have noted already that I take the assumption that our epistemic access to the world is limited to the deliverances of the sense organs—what I call the fourth dogma of empiricism—to be a mistake, but perhaps this seems an extremely dubious suggestion. What could be more a truism than Quine's insistence that "whatever evidence there is for science is sensory evidence . . . the stimulation of sensory receptors is all the evidence anyone has to go on, ultimately, in arriving at his picture of the world"?[9] What, that is, could I be denying? The fourth dogma, what I wish to deny, is what is left of empiricism once we have purged from it Quine's two and Davidson's third dogmas. These dogmas relate to the nature of experience itself: to accept the arguments of Quine and Davidson (which I relate in detail in chapter 3) is to deny that, prior to conceptual synthesis, sensation has qualities which can be experienced non-cognitively as such; it is to deny that (again, prior to cognitive interpretation) experience has what Rorty calls "raw feels" in terms of which we can access, or get some phenomenological handle on, our sensory experience. Quine and Davidson do not claim that our experience has *no* phenomenological content (that there is nothing it "feels like" to see that the grass is green); instead they argue that whatever experience may feel like, no stimulation of our sensory receptors warrants the name "experience" except that which is conceptually structured. Our sensory experience of the world is always already arranged by and in terms of our conceptual structures. Such arguments, presented as convincingly as they are, spell the end of the empiricist notion of "pure experience," of an autonomous stream of sensation with intrinsic and accessible qualities on the basis of which we judge the status of the physical world, back to which all our judgements can therefore be traced, and against which those judgements may be checked. The distinction this requires between our conceptual scheme, on the one hand, and experience, on the other, is simply not tenable. Concepts are not brought to bear on the stream of experience; our experience is—comes to us already—conceptually structured. The depth of this critique entices Davidson to claim that after we deny the third dogma there is nothing left to the empiricist position. But it is not so. It is true that the account of the *nature* of sensory experience we are left with would be quite alien to the classical empiricist.[10] But the notion remains that such experience as this is our only sort of contact with the world with cognitive and epistemic significance. It is this I wish to deny. Sensory experience may have a nature just as Quine and Davidson describe, but if so there is more to our epistemic contact with the world than sensory experience. This deserves some immediate explanation, and because I will much later in this work be following Aristotle's lead, I would like to offer this explanation in terms of Aristotle's theory of knowledge, in which, if I am right, the active body played an epistemic role irreducible to sensation.[11]

Aristotle distinguishes four ways (or kinds) of Being, and, likewise, four ways of Becoming, from the cooperation of which arises each successive state of our eternally existing, ever changing world. Together, Primary Substances (material particulars), Secondary Substances (forms), Accidents (accidental properties) and Universals (types and genera of accidents) compose all the entities of the world.[12] Aristotle situates his epistemic subject firmly within this single, composite realm of existence (unlike Plato, whose composite subject straddles two distinct ontic realms) and endows her not merely with the usual epistemic accoutrements (sensation and some form of Intuition[13]), but also stresses her concrete involvement in the world as important to her ability to know it. For Aristotle the human being is both caused by, and is a cause in, the world—she is involved not merely as a being, but also as a participant in becoming. Because of the prominence Aristotle gives to a science of Becoming as necessary to complete knowledge of the world, the practical, causal involvements of the human are, for Aristotle, important for her role as knower.[14] This emphasis on the importance of praxis is among the more important and unique aspects of Aristotle's philosophy, and it extends not just to ethics (where it must obviously be central) but to science and epistemology as well. One need only consider Aristotle's admission in the *Nicomachean Ethics* that practical activities can actually change one's perceptual relation to the world to see how deeply runs the importance of praxis to his philosophy.[15] Even his science understood itself to be as much practical and interactive as theoretical and (passively) observational.

This aspect of his work has its roots, I believe, in Aristotle's particular treatment of form as a metaphysical category. As is well known, Aristotle inherited from his teacher a strong belief that essence was to be found in form; to classify a thing by natural kind was precisely to classify it by form, and hence the importance of form as a category of Being. Less often emphasized is the role form plays as a category of Becoming, and the importance of formal cause as a fundamental, or essential, explanation of events. But it is precisely because of the connection between these two metaphysical roles of form that praxis emerges, for Aristotle, as an important mode of our relation to the world.

For substances do not merely possess a characteristic way of Being (essence) in virtue of their form, but also a characteristic way of Becoming. For each kind of thing, and therefore each substance, there are certain natural potentials, brought to the substance by form, which that substance will tend to actualize throughout its existence. These natural potentials range from the mundane tendency of physical objects to return to the earth, to the complex development of living organisms from fetus to adult. Although of course fundamental explanations of motion are for Aristotle always teleological, it is only in virtue of form that substances possess a *telos*. From the point of view of physics, then, formal and final explanations are importantly interrelated; to know the essence (form) of a thing is to know its natural tendencies, to see where it fits into the network of nature's motions.

In an important reversal of Plato's order of knowledge, (a consequence of Aristotle's insistence that forms exist only in virtue of their inherence in matter, i.e., insofar as matter and form together comprise substances) Aristotle maintains that the primary objects of knowledge are substances and that therefore form, rather than being available to immediate intuition, is an object of (scientific) discovery. It is, I believe, in this discovery of form that praxis plays its decisive role. For given the close ties between the essence (form) of substances and their natural tendencies—the principles of motion and rest contained in each subject in virtue of its form—one may come to know form by the discovery of these tendencies. And it is only in virtue of one's physical presence in the world and the capacity for practical interaction with substances which this entails that these natural tendencies, which exhibit themselves causally, may be discovered. It is my understanding that Aristotle's insistence on the metaphysical importance of Becoming (both to the nature of the world, and to the essence of individual substances) led him to see the importance to knowledge of interactive investigation as a mode of epistemic access to the world.[16] I think we can fruitfully understand Aristotelian science as positing a kind of natural ecology, being the totality of the interacting natural tendencies of substantial individuals. We are able to discover these tendencies precisely in virtue of our active presence in the world. The actions of substances affect us not just by stimulating our sensory receptors, but—by moving or resisting movement, changing or resisting change—substances react to, and alter, our behavior. It is my claim that this effect, too, has epistemic import.

I have no wish to provide any more detailed account of the role of praxis in Aristotle's "epistemology" or scientific method, and I will wait until chapter 4 before elaborating in detail my own adoption of this general picture. But as it is toward such a picture that I am arguing, it will be useful to bear in mind as we proceed. How the inclusion of comportment as itself a mode of epistemic access to the world supports the "aboutness" of knowledge in a way which sensation alone cannot will become clearer along the way.

1.3: Epistemology and Aboutness

The general shape of the problem which our acceptance of empiricism's fourth dogma presents for "aboutness" (and later, as we shall see, for intentionality more specifically) is easily seen. For once we restrict epistemic access to sensation—that is, to the causal impacts of the world on the sense organs—and thus purge sensation of empirical content—it becomes clear that for knowledge to be possible, the mind must fill the content gap, somehow making conceptual sense from causal force. This way of thinking (borrowed from John McDowell) is a new way of understanding an old problem: whereas Davidson's sensations need the mind because they have *no* content, insofar as the content accorded to sensation by earlier thinkers can be shown insufficient or suspect, the mind must be there, too, for the same purpose. To address only the familiar skeptical aspect of this concern, since the "feelings" or "intuitions" of bodily perception might be utterly individual (and surely provide no ground for suppositions of their universal quality) an attractive way to account for our shared (objective) conceptions of reality is to suppose that the *significance* of these sensations—both with respect to the object-properties thereby presented to the sensations, and to the identities and boundaries of the objects themselves—is a product of (on a Cartesian model) our *judgement* of their significance, or (on a Kantian model) their *synthesis* in accordance with shared rules and procedures. But although Kant, Descartes and others were surely correct to notice that sensation itself is not rich enough to provide the rules and cues by which we aggregate perceptions into objects and such, it must be admitted there is something ironic in the fact that the intersubjective validity which the synthetic model offers appears to be bought at the price of knowing the world itself; it is the very insertion of judgement between the "information" gathered from the world and "knowledge," intended to avoid skepticism, which in fact allows for truly radical doubt. For once our *access* to the world is thought to be restricted to sensation, to insist that knowledge is in fact the determined *significance* of this information is just to add the arbitrary to the fallible.

In the Cartesian case, the problem is the lack of epistemic evidence supporting one possible existential entailment over another as an appropriate inference from a sensory perception—it is impossible to decide what any particular experience means for the world. It is not so much that sensory reports may not indicate some truths about the world, but that it is impossible to decide which of the possible truths implied by a given sensory experience-type any particular *instance* of experience is to be interpreted as indicating. From a perception of a stop sign, for instance, should I conclude that there is a stop sign before me? But perhaps I am dreaming, or mis-seeing something *as* a stop sign; yet still one could conclude from this the truth of the general statement that stop signs exist (since I must have gotten the idea from somewhere). Yet, of course, unicorns do not exist: I have just combined the ideas of horse and horn. So perhaps I can conclude only that octagons and the color red exist? Or perhaps all that is needed is half of an octagon, doubled and combined. Or maybe eight short line segments would do, or for that matter, the individual points which make up the line segments. However clear our perception of that stop sign may be, in the absence of further evidence of unquestionable significance, any choice between the possibly infinite number of its potential existential entailments is, from the standpoint of reality, no better than arbitrary.[17]

The Kantian model does not eschew this arbitrariness, it merely displaces it onto the very conceptual matrix preconsciously arranging our intuitions into conscious experience. Why should we have just this set of concepts? Why should the world be *for us* just this way given that the answer "because the world *is* just this way" is disallowed by the idealist context? Kant, of course, read with some degree of charity, has some answers to these questions, and it may indeed be true that on Kant's full and complicated account of our epistemic relation to the empirical world, these problems do not arise so easily, and certainly not so immediately.[18] But in fact we have imported into contemporary theory the (undoubtedly true) claim that "percepts without concepts are blind,"[19] with neither the counterbalancing Kantian claim that concepts without intuitions are empty,[20] nor Kant's transcendental frame.[21] And given only this dictum we are no doubt forced to concede a certain degree of epistemic idealism, for if we are to know and understand the world, such understanding must come in terms anthropic enough to be accessible to the human mind. Given that our concepts are necessarily "human," rooted in the particular human needs, desires and interests informing our conceptual apparatus, our perspective on and interpretation of the world will likewise be specifically "human," which seems to imply a certain attenuation of objectivity. This impression is heightened when we consider the case of higher level scientific knowledge, for here it must be admitted that scientific observation, like everyday perception, is "theory-laden;" the significance of any perception, the very characterization of an observed phenomenon as an instance of some particular *type* of occurrence, will depend in large measure on the theoretical apparatus brought to bear in interpreting that perception.[22] Kuhn's reflections on the history of science, (on particular historical moments, as well as the movement of history as such) brought to light the epistemic significance of the competition of (apparently) incommensurable theoretical systems. Inevitably, a decision must be made between such competing systems, but if observations get their significance only as a result of their place in a particular theory, then a preference for a given theoretical frame cannot, it seems, be grounded in the claim that one better explains the "facts," for facts are artifacts, the result precisely of conceptually and theoretically interpreted observation. There are no neutral "facts" the explanation of which by two theories could be comparatively evaluated. Whatever our notion of a "better" theory amounts to, then, it is not the one which more closely corresponds to a world of facts. We seem left with the realization that the "better" theory is instead that which more adequately realizes our needs, desires and interests; our scientific access to the world leaves us no closer to objectivity than does our conceptual access.[23] For if, instead, the terms in which we fashion our understanding were to be judged by their relative appropriateness to the world—their (approximate) truth—it seems that we would have to have some way of getting "outside" our conceptually mediated perceptual access to the world; and this is just what we cannot do.[24] Percepts without concepts, after all, are blind.[25]

The problem which emerges here is *not* the narrowly skeptical one that the world might be other than our concepts make it seem, but is rather that by making concepts alone determine the content of knowledge, sensation becomes empirically superfluous. It is difficult to see, on the theory-laden perceptual model of our access to the world, how experience can operate as a "test" of theoretical adequacy (and be, therefore, a guide for accurate knowledge). For the operative ideal of such testing would be the absence of experiential anomalies. But if an experience is transparent enough to be conceptualized as an anomaly, this implies that the experience has been rendered intelligible by some conceptual apparatus, that is, that it has been interpreted by some "theory," in which it does *not* appear as an anomaly. As such it fails to be a test of the theory$_1$ in question. The appearance of such an "anomaly" can at most imply that theory$_1$ is simply not the theory which is conceptually active for the perceiver, that is, that perceptions are not theory$_1$-laden. Furthermore, we are left with the same dilemma for the "more basic" theory$_2$ (or general conceptual schema) in which the "anomaly" appeared. Either the experience is intelligible according to the criteria of that theory, in which case it is not an anomaly, or the experience is not conceptually available given the resources of the perception-informing theory. In the latter case, the experience must fail to be integrable into a coherent account of the world, which is just to say that it fails to appear *as* an experience of some world (even one different from the one theoretically expected). As such, the "experience" cannot offer grounds for modifying a theory in light of anomalous evidence.[26] But if experience cannot play this epistemic role, then there is no sense in calling it a mode of *epistemic* access to the world: it is reduced to a conduit through which nothing of epistemic import comes. If this is so, then our knowledge must stop short of the world, and this surely threatens the notion that our knowledge is *about* the world, and with it the hope that our knowledge is true of the world.

If we do not wish to accept the above consequences, we must admit *either* that some perceptions are not theory-laden (are not blind even though conceptionless,[27] which would be a return to a more classical-minded empiricism) *or* that there is some nonperceptual (nonsensual) mode of epistemic access to the world, a mode of access which is not theoretically mediated in its contact with the world. I urge embracing the second lemma, arguing that the active body is epistemically "open" to the world; the body is a mode of access to the world with the resources to reveal theoretical anomalies, for concepts are receptive to the world in activity in ways they are not in perception. No matter what our conceptual organization of the world leads us to believe about ("see in?") a particular physical system, actual interaction with that system can provide cause for rethinking these beliefs.[28]

There are simpler reasons for wanting a richer account of our epistemic encounter with the world. On a sensual model of epistemic access we are forced to account for our knowledge of *all* the aspects of the known-world on the perceptual model. We ought not resist the suspicion that this is asking one model to explain too much. To cite just one example of this overextension, it seems especially unpromising to treat such aspects of the world as solidity or weight on the sensual model; surely it is more plausible on the face of it to treat our knowledge of the solid in terms of its relation (resistance) to our behavior, rather than of some "sensation" of solidity. Still, these are just suggestions (although I hope tantalizing ones), and they can be vindicated only in the course of this book as a whole.

1.4: Epistemology, Intentionality, and Descriptions

Contemporary epistemologists are not, of course, unaware that knowledge ought to have an object. But in a striking example of the division of intellectual labor, made possible by philosophy's "linguistic turn," the account of the connection between knowledge and its particular objects—the account, in short, of intentionality—has been largely guided by semantics. This is not, in the end, particularly surprising, since much contemporary epistemology seems guided by the insight that the paradigm case of knowledge is descriptive, where a description is something (perhaps a linguistically structured mental content or propositional attitude) derived from or otherwise related to sensual experience.[29] Thus it will be useful to make a few remarks about the philosophical attraction of descriptions.

Ian Hacking offers the following as a kind of explanatory summary of contemporary philosophy of language:

> The most naturally occurring analysis concerns sentences of subject-

predicate form: "The marigold is orange." Draw up a square like this:

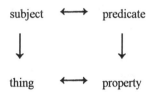

We assert that S is P. S refers to a thing, reference goes down the left side of the square. P indicates a property, this goes down the right side of the square. If the two downward strokes succeed, and we successfully refer to a thing and indicate a property, then we can start saying something.[30]

Unfortunately, Hacking notes, this picture is not faithful to our phenomenological experience of objects or of our referential access to them. It is not at all clear that we separately confront objects and their properties as indicated by the separation of "things" and "properties" on the bottom world-line.

What we see is more like this, except in color . . . . If we are faithful to our experience we have to replace our square by a triangle:

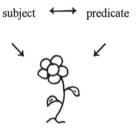

This diagram is much more mysterious, and raises two problems, first, how do the two downward arrows work in different ways, and secondly—the pressing problem for our general grammarians—what is the nature of the copula which rejoins what is separate.[31]

Part of what interests me in Hacking's commonsense rendering of the problems of philosophy of language is that, as in the above diagram, where we have two aspects of reality to which we want (referential) access, it is natural to posit one mode of access to each—a downstroke for the "thing" and another for the "property." But of course a great deal of contemporary philosophy of language is precisely a repudiation of this model. As Hacking notes, Wittgenstein's early work was motivated by the intractability of the grammatical problems presented by a world of two-aspect things, which problems he dissolves by positing a world of *facts*— replacing things with something like atomic linguistic items. Likewise, as it is easy to see on Hacking's model, the assumption that we can have only *one* mode of access to the world of objects makes necessary a kind of descriptive mentalism, whereby we access the sensually available properties of things (the descriptions under which they fall) and, in virtue of certain capacities of mind, unite these properties into our conceptions of objects.

But this puts us in a position to see the attraction of descriptions as an epistemological (theoretical) tool. They are ideal as a way of characterizing our access to the world, especially if we understand that access primarily in terms of the sensual properties of things (even the truth-producing models of perception revolve around producing beliefs containing reference to sensibly available object-properties), since those properties represent our means and terms of *description*. So as an epistemic item, a description seems to represent well the form of our objective knowledge.

  Descriptions qua linguistic items possess important features too, among which is their apparently built-in intentionality: descriptions seem naturally to be *about* things. By their very structure they seem to point toward an object, and thus they have had an important place in many explanations of our referential abilities. If we need referential access to objects, but have epistemic access only to properties, we need some tool by which we can reach to objects with the knowledge we have; linguistic descriptions are meant to provide this service, since they seem to indicate the *object* descriptively characterized.

This allegiance, of course, has its ontological consequences, which are, I think, well illustrated by the "cluster theory" of objecthood. Generally related to inquiries into the meaning of proper names (that in virtue of which, it seems, we identify and/or refer to particular objects) the gist of cluster theories is well illustrated by Bertrand Russell:

> it is natural . . . to suppose that a proper name which can be used significantly stands for a single entity; we suppose that there is a certain more or less persistent being called 'Socrates' because the same name is applied to a series of occurrences which we are led to regard as appearances of this one being.[32]

It is well known that Russell held such proper names to be about that single entity (the myriad appearances of which we were led to associate with a single being) in virtue of being, or signifying, definite descriptions of the form "the so-and-so."[33] Thus we are held to know and access the object so named (described) in terms of its definable features, features which we come to associate with the same object. Of course, the complexity of Russell's account precludes him from being characterized without qualification as a "cluster theorist," so I turn instead to the more easily caricatured account of John Searle.

Searle takes his leave from Frege, and builds a theory of naming from the idea that, although names themselves are not descriptive, nevertheless they refer to their objects by means of descriptions. Names, as he puts it, "function not as descriptions, but as pegs on which to hang descriptions."[34] In other words, a proper name is associated with some finite but indefinite list of descriptions thought to characterize the object in question, and our ability to use a name to refer to said object depends upon possessing a conception of the referent (associated with the name) consisting of at least one appropriate descriptive attribution. (In the case where we have more than one description associated with the name, something like the decided majority must be appropriate.)

Indeed, it seems a natural consequence of the (reasonable) claim that our epistemic relation to objects is characterized by (and our capacity to individuate and sort is dependent upon) access to their perceptually available features, that reference would also depend upon and be determined by this knowledge. The known properties of objects would constitute a kind of list of the descriptions which characterize the object and allow us to identify it as such. And given the apparent "intentionality" of descriptions themselves, their role in reference seems clear.

But as I will argue later (in both chapters 2 and 4), although descriptions play an important role in our knowledge, and although knowledge of (or, more precisely, the possibility of epistemic access to) objects is indeed a condition for reference, descriptions can account for reference and intentionality *only* in virtue of our capacity to *individuate* objects, and descriptive pointing sheds no light on this latter capacity. For it looks as though, in order to create a list of properties which characterize an object, one would have to have *already* identified the object, for without pre-existing conceptions of its boundaries it would not be possible to differentiate between the characteristics belonging to the object, and those belonging to its background or surroundings. Without such a conception of the object and its boundaries it would be difficult to know when to end the list of descriptions of spatially contiguous color-shape-texture areas within one's sensory field, i.e., when one had reached the edges of the object rather than some qualitatively homogenous part of the object, or gone beyond the object to describe some other spatially related entity. Sensation alone does not provide us with the resources to make nonarbitrary determinations as to the boundaries of a given material particular. This invites a proposal to the effect that we "know" objects before we sense them: but insofar as this implies knowledge of bare particulars, we must reject the suggestion, for what, of such particulars could we be said to know?[35]

Such concerns lead easily to full-blown skeptical worries, for skepticism takes root in the alleged gap between the metaphysical nature of the object and our epistemic perspective on it; we are faced with the question of whether to reduce the criteria of identity for objects to their epistemically available criteria of identification (and thus make objects as such ideal, artificial, but knowable), or to maintain the distinction and imagine we are cut off from metaphysical reality (and thus make objects real and unknowable). It seems implausible and *ad hoc* to define an object as a collection of sensations, descriptions, or sensation-producing potentials,[36] but just as problematic to imagine that we have the kind of contentless knowledge which embracing the other lemma seems to necessitate. For consider the most common competing alternative, born of Hilary Putnam and Saul Kripke's thorough critique of the Searlean account of naming on precisely the grounds that descriptions cannot determine reference, as is apparently required if they are to play a central role in intentionality. According to Kripke and Putnam, we should treat terms (e.g., natural kind terms) as rigid designators which refer in virtue of a sortal essence possessed by all members of a given class.[37] But, as I will argue in greater detail in sections 2.3 and 4.1, if this is to imply that reference is not guided or limited by knowledge, then we face the potential of a skeptical gap between intention (what one is thinking about) and reference (what one is, in fact, referring to). For if we have no epistemic access to the features of objects which determine reference (the "sortal essence"), then we may not be referring with our use of "lemon" to that to which we take ourselves to be referring. Indeed, this is a consequence which Putnam himself exploits in defense of his later formulated "Internal Realism."[38] Yet if we *do* have some epistemic access to reference-determining features of objects, then by the fourth dogma it can only be in virtue of perceptual access to sensual properties.[39]

1.5: Knowledge and the Body
   We can see how the fourth dogma forces us to oscillate between skepticism and idealism, with "aboutness" and intentionality in constant danger. If we insist, reasonably, that in order to refer to or think about some particular we must know *which* object we intend, we seem pushed towards idealism, because according to the constraints of the fourth dogma, in the absence of any other available criterion defining an object's identity, the object must be defined in terms of our mental arrangement of sensual material. Yet this very acceptance threatens our claim to know *about* some particular in the world, for if the object is by definition a (synthetic) sensual construction it is not necessarily in, nor even derived from, the world at all. In such a case we lack epistemic *openness* to the ontic structure of the world. If, however, we relax the "knowing which" constraint to allow reference to proceed without knowledge—perhaps determined by causal connections with some or another feature of reality—then we face the problem of skepticism. Since we have access to only sensual properties, and since it is not by these properties that reference is determined, we do not have epistemic *access* to that to which we ostensibly refer. In both cases, the object of reference is providing no epistemic friction or guiding influence for our conception of it—which surely makes doubtful, or at least untestable, the claim that there *is* some veridical connection between our thoughts and their supposed objects.

Without *epistemic* connection, the claim to intentional or referential connection looks tenuous.

The general idea is simply this: adherence to the fourth dogma of empiricism makes it difficult to avoid the need to make mind the originator of empirical content; and to do this is to make what we might call the phenomenological order—the "world" of our "experience"—epistemically closed. Content can be traced to mind and no further; our cognitive insides cannot be related to the external, and thus we cannot be supposed to be in touch with the world . . . unless, of course, we make of the world an internal artifact.

Of course, I cannot claim that the committed epistemologist lacks the resources with which to deal with these very general problems; such considerations have been the source of numerous interesting and complicated contemporary debates. But for me, at least, the frequency and familiarity of these choices and debates illuminate the need for an alternative account, for it raises the suspicion that there is some feature of our picture of the nature and position of the knowing subject which make the dilemmas inevitable. What I hope to show is that attention paid to the epistemic importance of the active body is well rewarded; dilemmas like the above are compelling only when we accept the fourth dogma of empiricism, that we have but one, sensual mode of epistemic access to the world. I will argue that the active body offers a mode of epistemic access to the world which compliments but is not reducible to the mode of access offered by sensation.[40]

In overview, I will be arguing as follows: chapter 2 looks to establish, through a detailed look at Frege's semantics (the origin of the body of work which takes reference to depend upon knowledge) that when intentional/referential connection rests entirely on sensually gathered knowledge, one is indeed caught in the skepticism-idealism dilemma, but that supposing reference to proceed independent of knowledge is an unacceptable alternative. However, I will argue that a Fregean theory of intentionality/reference might be workable (within the metaphysical constraints I have set) if we give up on the idea that our *individuation* of particulars depends upon sensual access to them, and restrict its application only to the *discrimination* between pre-defined and apprehended particulars. This, of course, leaves a deferred project of ontology at the core of Frege's semantics: how, and in virtue of what epistemic faculties, do we manage the individuation of material particulars upon which this otherwise compelling semantic theory seems to depend?

I will not address this question explicitly until chapter 4, and proceed instead, in chapter 3, to argue for the connection between "aboutness" and "epistemic openness" with a detailed look at some central texts in American pragmatism, and its inheritance in contemporary American thought. In essence, I argue that we cannot give an account of our intentional connection with a metaphysically real world unless we can posit a form of epistemic openness to that world sufficient to allow for the world to limit and guide our conceptions of it. But sensation-based models of our epistemic access to the world cannot allow the necessary degree of openness without violating the set metaphysical constraints, and jeopardizing the intentional connection we mean to preserve. It is meant to be clear by the end of chapter 3 that we must look beyond the fourth dogma of empiricism if we are to provide an acceptable account of our epistemic access to the world.

Chapter 4 begins by combining the arguments of the previous two chapters with a critical look at Locke's empiricism. This sets the stage (and fixes a convenient working vocabulary) for the proposed solutions. Thus, in the service of Frege's deferred project of ontology, I will argue that our individuation of material particulars is primarily attributable to our physical presence and active interference in the world (rather than to our sensual reception of it) and that the empirical concepts most responsible for the synthesis of sensual information are themselves open to the world in virtue of this embodied activity and interference in the natural order. This is to say that the form of our epistemic receptivity to the world is behavioral—comportmental, to use a term with the proper resonances — and this comportmental receptivity is primarily responsible for influencing the formation and development of our empirical concepts. Thus it allows us to account for both the general "aboutness" of our knowledge and for some central conditions of intentionality.

In its overall scope, my argument is both empirical and transcendental. It is, I will claim, an empirical fact that our concepts are open to the world in activity, that they are acquired in the course of, and get their particular synthetic significance (play the role in perception which they do) because of, our active and practical engagement with the world. But I will also make the transcendental claim that this active engagement with the world, and the conceptual openness which it allows, is a condition of intentionality and "aboutness;" without this openness we could not be "in touch with" the world.[41]

Allow me, now, to end these rather long introductory remarks with a caution: my position is *not* the familiar pragmatic one. To get us back in touch with the world it will not be enough to admit—as many have already done—that, given the location of our senses on a mobile body, how we move, where we go, and how we are positioned will have a great deal to do with what we experience and thus what we come to believe (or know). For as long as we leave in place an epistemic/cognitive psychology which includes the fourth dogma of empiricism, we will be naturally led to a picture of epistemic confinement, forced, as it were, to acknowledge our distance from an essentially unreachable, unknowable world. I am arguing not just that the body is important to empirical thinking and knowledge, but that it is essential, and essential in a rather peculiar way. I myself do not know what the implications of this view are, or even whether I should be content to identify myself with the full extent of them. But I hope that they will prove unfamiliar enough to be interesting, and yet reasonable enough to be worth pursuing.

Notes

1.   Crispin Wright, "Realism, Antirealism, Irrealism, Quasirealism," in *Midwest Studies in Philosophy XII: Realism and Antirealism*, Peter A. French, Theodore E. Uehling, and Howard K. Wettstein, eds. (Minneapolis, MN: University of Minnesota, 1988), 25.
2.   Lest this assumption seem to beg the question against the skeptic, let me make one simple point: skepticism possesses the power it does precisely because it can take common-sense assumptions as the starting point for an apparently inexorable series of arguments which reveal the skeptical implications of those same commonsense assumptions. To deny at the outset our physical existence in a real, finely structured world is simply to cede ground to the skeptic without argument. This I do not intend to do, but nor do I intend to address directly the skeptic, or skeptical arguments. Whatever use the considerations of this work may have against the skeptic (and I will make this explicit in a later chapter) will be in virtue of having provided an alternative to the common sense of our epistemic position.
3.   With, that is, the general "cognitive psychology" which these assumptions represent. Of course, the question naturally arises: "Which epistemic theories accept these objectionable premises, making them (theoretically) vulnerable to your critiques?" This question is in part left to the reader. I do not offer a survey of current theory. Not only are such surveys (of necessity) boring and pedantic, but they cannot hope to do justice to the particulars of the material under review, and they run the risk of distracting the reader from the central argument of the text.   Instead, I treat in some detail the work of Donald Davidson, John McDowell, Hilary Putnam, and Richard Rorty. My discussion of this work, as with that of such figures as Frege and Locke, is meant to illuminate the aspects of their particular theories which make them vulnerable to the critique I offer.
4.   Why semantics? Because in this subfield of analytic philosophy are to be found some of the most sensitive treatments of the "aboutness" of mental states, and of the conditions (epistemic and others) which are necesssary for our intentional connection to the (particulars in the) world. And why pragmatism? Because the pragmatists were centrally concerned with (among other things) how (or whether) the world could provide "epistemic friction" sufficient to guide our conceptions of it closer to veridicality. I intend to identify adherence to the assumptions mentioned above as a common root cause of the difficulty in accounting for "aboutness" in semantics and "epistemic friction" in pragmatism, and I will propose the beginnings of (or a path towards) a solution to these long-standing difficulties.
5.   This argument owes much to John McDowell's *Mind and World* (Cambridge, MA: Harvard University Press, 1994).
6.   It should be noted at the outset that I utilize a very thin conception of the body (and embodiment) here, thus: the body is an object, situated in the physical world as the locus of causal interaction with the world, our awareness of the bounds of which object is a necessary condition for (our) subjectivity and agency. The width of this conception is one practical upshot of the form this book takes: the concept of the body is itself a shell. This is not because I think this is all which there is to, or can be known about the body, but

rather because I wish elaborations of the concept of the body to serve as elaborations (that is, complications without being refutations) of my general thesis here. As minimal as this conception of the body seems to me, it has been objected that it does in fact assume too much. In one instance, after making some of the arguments of this work at a Comparative Literature conference, it was objected that I left no room for the fact that the body is ultimately a fantasy, that one might just as well choose a virtual body (choose, for instance, to live in cyberspace). Here my insistence on the physicality of the body unnecessarily and unfairly biases my arguments in favor of metaphysical realism. Whatever the merits of such objections (and I take it certain subtle and challenging versions of this thesis are to be found in such work as Judith Butler's *Bodies that Matter* [New York: Routledge, 1993], addressed by many of the contributors to *The Incorporated Self*, M. O'Donovan-Anderson, ed. [Lanham, MD: Rowman & Littlefield, 1996], and with which I hope personally to deal more carefully in later work), I must admit I felt somewhat vindicated when my interlocutor boarded the bus for her trip back to her home university; after all, e-mail would have been much faster and less expensive!

7.   It has long been understood that the sensual model of epistemic access has great trouble accounting for the world's bearing on belief; this suggests the possibility of radical incongruence between belief and the world which grounds the skeptical position. It is hardly far-fetched to see escape from these implications as a motivating force behind idealism (as it certainly was for Kant). I am taking direct issue with the "cognitive psychology" of epistemic sensualism and, to this degree, am taking direct issue with one of the assumptions which supports the most common versions of skepticism. (This is, I hope, a weakening of, but is not the same as a refutation of, skepticism.) But it is only insofar as the (apparent) epistemic distance between mind and world implied by the assumptions I question is a *motivation* for idealism that I am also questioning idealism itself. That is, I am not arguing against idealism, but one of idealism's motivations; I am not refuting skepticism, but one of skepticism's grounds. And regardless of the success of these arguments in convincing the reader that I am justified in doing so, I am explicitly working under the auspices of realism.

8.   Just as the myriad discrete epistemological theories defy any attempt to address them adequately as such.

9.   W. V. O. Quine, *Ontological Relativity* (New York: Columbia University, 1969), 75. Interestingly, even Michael Williams, who in *Unnatural Doubts* is at pains to deny the skeptic his apparent due, wants to retain the "truistic character" of the claim that knowledge depends on the senses (see pages 68-9).

10.   It is very hard to *say* what the post-empiricist notion of experience is: perhaps it is the end of a process which begins with the stimulation by the physical world of our sensory receptors, in which the stimulation calls into operation some set or series of empirical concepts, and which calls upon us to judge of some feature of our surroundings. The "shape" of our consciousness of the world resulting from some such sequence, a conscious state populated by types and divisions, categories, colors, individuals, and kinds, textures and groups, wholes and parts, is what we mean by sensory experience. This is not a happy formulation, but I will not tax its details. My inquiry is not into the nature of sensory experience, but into the foundation of its content and "aboutness." Thus I can provisionally endorse even such an unhappy formulation as the above as a way of fixing my reference to "sensory experience."

11. The common claim that activity is epistemically important because it increases one's sensory exposure to the world advocates an epistemic role for the body which is *reducible* to sensation.

12. There is some question as to whether Secondary Substances and Genera are in fact distinct categories of Being; this dispute is immaterial to the argument of this chapter.

13. It does not matter to the argument of this section if Stanley Rosen is correct in his assertion that there is not, in Aristotle, a faculty of Categorical Intuition.

14. Harry Prosch writes that "Aristotle appears to have been the first philosopher who attempted to rescue the world of becoming . . . for philosophic purposes. He developed a theory of empirical knowledge and inquiry which was designed to find something knowable in the world of change, namely its stable 'processes.'" *The Genesis of Twentieth Century Philosophy* (Garden City, NJ: Doubleday, 1964), 294.

In what follows I will explain, but not defend, my reading of the significance of praxis in Aristotle's epistemology. No doubt the reading deserves some defense, but I do not wish to mar the simplicity of this introduction given that the defensibility of my reading is a matter of complete indifference to its function as an introduction to the issues to be dealt with in the following chapters.

15. See, for instance, [*Nicomachean Ethics* 1114b] "But someone may say, 'Everyone aims at the apparent good, and does not control how it appears; on the contrary his character controls how the end appears to him.'" Character, of course, is acquired primarily through habit (practical activity).

I think it is clear that for Aristotle's ethical system to work, perception—and with it the descriptions under which objects and events appear—must be sensitive to character and habituation. This is why he does not simply dismiss the objection as unfounded, but instead insists that actions will still "depend on the agent" insofar as character does.

16. In "Thought and Action in Aristotle," Elizabeth Anscombe develops doing as a way of knowing, as it appears in Aristotle's ethics. Beginning from considerations about deliberation, and its relation to the practical syllogism, she argues that for Aristotle, deliberation is itself a kind of action—reason flows naturally into activity. It is a short step to posit the possibility of activity with its own (noncognitive) content, sufficiently robust to accept characterization in terms of truth and falsity, and it looks as though Anscombe has something of the sort in mind with her analysis of "practical truth."

"It is practical truth when the judgements involved in the formation of the choice leading to the action are all true, but the practical truth is not the truth of those *judgements*. For it is clearly that 'truth in agreement with right desire' which is spoken of as the good working out of practical intelligence, that is brought about—i.e. is made true—by action . . . . The notion of truth or falsehood in action would quite generally be countered with the objection that 'true' and 'false' are senseless predicates as applied to what is done. If I am right there is philosophy to the contrary in Aristotle." *From Parmenides to Wittgenstein* (Minneapolis, MN: University of Minnesota Press, 1981), 77.

17. This is to say, we may call on considerations like coherence with past judgements, but we will have lost anything like the objectivity (the overcoming of fallibility) which judgement or synthesis was meant to buy us.

For a more complete account of the nature of Cartesian skepticism see my "Certainty, Doubt and Truth: on the Nature and Scope of Methodological Doubt in Descartes' *Meditations*."

18. Both Karsten Harries and Susan Shell have reminded us that there are resources in Kant's philosophy that take us closer to embodiment and interactive access to reality than the above would indicate. See especially Susan Shell *The Embodiment of Reason*, Michael Friedman *Kant and the Exact Sciences*, and Kant's own *Opus Postumum*. (I hope to have the opportunity to explore Kant's notion of our embodied access to reality in a planned work on [primarily] the realism of Emerson's "Nature" and Heidegger's "The Question Concerning Technology.")

19. In the most extreme forms of mentalism, the information gathered by the senses (or however) is denied epistemic significance at any stage prior to mental processing. Thus Dretske writes that in order to make clear the "distinctive role of sense experience" in the acquisition of knowledge, "it will be necessary to examine the way information can be delivered and made available to the cognitive centers without itself qualifying for cognitive attributes—without itself having the kind of structure associated with knowledge." ("Sensation and Perception" in *Perceptual Knowledge* Jonathan Dancy, ed. [Oxford: Oxford University, 1988], 145.) In fairness to the field, it should be noted that Dretske is countering what he takes to be an undesirable tendency to endow perceptual systems with cognitive or quasi-cognitive abilities.

For my part, although I think that attempts to understand perception as itself an active faculty which has a constructive effect on the information it conveys can only be for the good, I find it instructive that these attempts tend to take the form of the expansion of cognition to include perception. Here again, our allegiance to the idea that the process of organizing (and otherwise making significant) perceptual information is the distinctive mark of *mental* activity continues unchallenged.

The problem, as already noted, is that if sensation provides our only epistemic conduit with the world, and if it cannot be thought of as providing reliable, or even significant content, then the knowledge we come to have cannot be said to be knowledge *of* the world. Its origin is in an important sense mental, and we are therefore faced with the apparent necessity of idealism.

20. In part this is because in contemporary empiricist language these "intuitions" would be "sensations," and it is precisely against the notion or utility of contentful sensations that the neo-Kantian argument is thought to be decisive (as presented, say, in the arguments of Quine or Sellars against the myth of the given).

21. Davidson can be understood as trying to replace this transcendental frame with language, so as to avoid the worst relativisms which result from the de-transcendentalized neo-Kantianism which is so apparently popular. I discuss the limitations of this approach in chapter 3.

22. This insight is borne out to some degree by the findings of contemporary cognitive science; we apparently engage in a great deal of "top-down" processing of sensual stimulation to determine its significance.

23. This position is well known from such works as *Ways of Worldmaking* by Nelson Goodman, *Objectivity, Relativism and Truth* by Richard Rorty (see especially "Solidarity or Objectivity") and *Reason, Truth and History* by Hilary Putnam. Much of the work was spurred by Thomas Kuhn's famous book *The Structure of Scientific Revolutions*. The similar argument denying the objectivity of the senses in everyday contexts is made quite thoroughly and convincingly by Kathleen Akins in "Of Sensory Systems and the 'Aboutness' of Mental States," *Journal of Philosophy* 43, no.7 (July, 1996): 337-72.

24. This position has both epistemic and political dimensions, for once upon reflection we deny that the accuracy of our conception of the world (theoretical or otherwise) can be accounted for in terms of an epistemically unproblematic set of perceptions, we are left free to proclaim the unavailability (and hence irrelevance) of the actual world. Having thus shrugged off the burden of reality, we may arrogate to ourselves the unhindered right to project onto the world whatever ideologically charged conceptual hierarchy of need we (a "we" bound by politico-epistemic solidarity) see fit, under the guise of constructive interpretation. But as works like *Dialectic of Enlightenment* (Theodor Adorno and Max Horkheimer) and *The Question Concerning Technology* (Martin Heidegger) make clear, such ways of non-world making inevitably lead to ways of world un-making. In addition to whatever epistemic motivations the preservation of truth claims might provide one, the high cost of theoretical self-absorption, both to the world and its denizens, provides political motivation for resisting the idealist argument. But how can we get beyond the limits of our own illusions?

25. For a nice take on the form and significance of this dilemma, see John McDowell's *Mind and World*, especially Lecture I.

26. This is, of course, just a version of Donald Davidson's argument in "On the Very Idea of a Conceptual Scheme": no verificationist-idealist theory of language or conceptual-structure can be challenged by anomalies, for nothing can simultaneously count as an anomaly and a language.

27. But I should be clear at the outset that I do not intend to deny the validity of the Kantian dictum regarding the conceptual significance of perception; as I argue in chapter 3 a "constructivist" or "foundationalist" approach leaves us no closer to knowledge, certainty, and truth than does the Kantian-holistic one.

28. On the cartoonish, practical joke side, you might be surprised when trying to lift a beer mug which turns out to be bolted to the bar, or when trying to follow Bugs Bunny through the tunnel he's painted on the cliff. When it comes to confronting bodily our perceptual illusions, we are all Wile E. Coyote.

29. Although it is certainly true that sense-*data* theories of perception are currently out of favor, perhaps for good, the most common way to analyze our epistemic access to the world is nevertheless in terms of perception, and perception in terms of sensation. Thus it is possible for Jonathan Dancy to remark, as an introduction to his volume on perception, that "Perceptual knowledge is the sort of knowledge that we get about the things around us by looking at them, feeling them, tasting them, and so on. . . . And it is largely by using our senses that we come to know anything about the world we live in. So the senses normally give us knowledge and that knowledge occupies a privileged position since it is by means of it that we are able to come to know other things." Dancy, *Perceptual Knowledge*, 1.

It is clear from context that the "other things" we can know are meant to be the product of mental activities like inference. Not all of our knowledge is directly about the world (since we have reason), but perception is our only mode of access to the world, and it is that upon which all of our knowledge which *is* about the world is based.

30. Ian Hacking, *Why Does Language Matter to Philosophy?* (Cambridge: Cambridge University, 1975), 86-7.

31. Hacking, *Why Does Language Matter to Philosophy?*, 87.

32. Bertrand Russell, "Logical Atomism," in *Logic and Knowledge* (London: Unwin Hyman, 1956), 331.

33. A clarification is in order. In, for instance "Descriptions," Russell distinguishes between being named and being described. But he soon restricts reference by "naming" to "logically proper names" like "here," and "now." Usually reference occurs by description, hidden or not, in which a person is known by some describable attribute which he possesses. Of course, Russell's insistence that logically proper names do not refer as such seems to indicate the possession of a nondescriptive mode of referential access to the world. See section 4.21 for Evans' important exploitation of Russell's analysis.
34. John Searle, "Proper Names," in *The Philosophy of Language*, A. P. Martinich, ed. (Oxford: Oxford University, 1985), 273.
35. Paul Ricoeur writes: "There is no science which does not make use of tacit presuppositions which are neither dependent on perceptions nor anticipations of objective knowledge and which bring into play a philosophical conception of nature, of human experience, and of man's position in the world. . . . This would then designate something other than the perceptive level of the knowledge of nature . . . [and] embrace a grasp of Being directly akin to perception but furnishing an implicit representation of the intelligible order..." *Main Trends in Philosophy* (New York: Holmes & Meier, 1978), 116-7.
36. When Locke or Hume suggested such "metaphysical" definitions it was precisely to close the gap in which (further) skepticism could take root.
37. See, e.g., Saul Kripke in *Naming and Necessity* and Hilary Putnam in, e.g., "The Meaning of 'Meaning.'"
38. See, in particular, *Reason, Truth and History* chapter 2, on the inscrutability of reference.
39. This is not entirely surprising, for the main thrust of the Putnam-Kripke criticism is directed against the notion that sortal concepts can consist of lists of *necessary* and *sufficient* features. The notion that a "weighted-most" of associated features helps determine the identity or class of an object is not directly challenged; surely (it might be argued) it would be foolish to deny that our capacity to recognize and sort particulars depends largely on their perceptually-available properties, even if we concede that our capacity to bring concepts to bear on individuals depends more on "family resemblance" notions than on necessary and sufficient conditions for identity. Thus the attachment to understanding our epistemic relation to the objects of the world in terms of perceptually available properties is not threatened by Putnam or Kripke. Indeed, Putnam notes, apparently with approval: "It is plausible, in terms of present-day brain research, work on artificial intelligence, and so on, that the brain contains devices for recognizing patterns (or, more generally, 'functions of observable properties')." *Representation and Reality*, (Cambridge, MA: MIT, 1988), 44.
   Putnam clearly accepts the epistemic role of descriptions without accepting their semantic role (denying that they give "meaning"). See Putnam, *Representation and Reality*, 38.
40. Understanding the body this way can provide us with the theoretical resources to begin to understand our contact with the world, and the structure of our experience, as more firmly grounded and metaphysically constrained than now-canonical philosophies allow. The active body—in the context of a pragmatism which takes behavior seriously, avoiding the contemporary tendency to analyze actions in terms of a temporally sequenced series of ("haptic") perceptions following upon an intention—may provide an Archimedean point, granting us the leverage to question the propriety of conceptual schema on grounds other than the ones they themselves construct. For it needs always to be remembered that

theories are never only aids to comprehension, but are also guides for action. Only if we treat awareness of our own behavior on the same sensual model as awareness of the world will we fail to see that activity may provide tests of theoretical adequacy not available to perception, because on this assumption we will be forced to treat scientific justification in terms of whether or not the expected (theoretically predicted) observations occur in response to some practical intervention: but of course these observations provide no more "objective" access to the world than those carried out in other contexts. The purported advantage of the experimental situation is that we can elicit such observations at will, in a controlled environment. But see Ian Hacking's *Representing and Intervening* for an excellent discussion, and critique, of this epistemology of science. A more directly relevant discussion, and one somewhat more detailed than the part which appears in chapter 4, may be had in my "Science and Things: On Scientific Method as Embodied Access to the World" in *The Incorporated Self.*

41.   As you will recall, I take much of the importance of retaining "aboutness" to derive from the fact that without "aboutness" there can be no truth. "Aboutness," epistemic access and openness share certain necessary conditions, in the absence of which truth is either beyond verification, or obtains *ex hypothesi.* Neither of these result in acceptable versions of truth.

2
The Deferred Project of Ontology in Frege's Semantics

2.1: A Brief Overview of Frege's Philosophy[1]

This chapter is concerned primarily with the theories of language of Gottlob Frege, but like Michael Dummett's extensive writings, my brief treatment of Frege is meant also to address some of the most central concerns of philosophy, and the ways in which these concerns are addressed by contemporary analytic theorists. Perhaps the most enduring of Frege's contributions was his analysis of the aboutness of language in terms of the cognitive content associated with understanding the semantic value of names and sentences; and rightly so, for there is something importantly correct to this position. What *makes* a thought or statement about some particular object is no doubt deeply connected with the capacity of any agent who grasps the thought (understands a statement) to thereby pick out the object meant. But I will argue that Frege's position, and by extension any position which accepts its assumptions, leaves undone a project essential to a complete understanding of our knowledge of the world and its objects: Frege's philosophy can at most explain that whereby we can differentiate particular objects in a field of objects already given as such; the capacities which allow objects to be so given are not thereby understood.

As I have hinted, this is not a problem isolated to Frege's thought. Indeed, as I understand it here the problem is a symptom of the ongoing debate between realism and idealism. Realism demands a level of objectivity where we are not the sole determinants of the shape of reality; that Frege's theory rests on the assumption of such a realm of independent objects is in this sense an unsurprising consequence of his realism. But Frege is no simple, doctrinaire realist, and there is room in his work for idealist tendencies. To the degree to which this is true, Frege can be seen as a symbol of the contemporary treatment of these philosophical themes; reactions to, and interpretations of Frege's philosophy are also thereby reactions to realism, idealism, or both.

Central to this age-old division is a disagreement over the nature of truth, its requirements and entailments. It is no different here. One of the reasons I begin with Frege is that I find something worth preserving in his (realist) conception of truth; the question which is constantly in the background of this work is how, whether, and to what degree one can accept the powerful critiques of such (idealist) figures as Wittgenstein without giving up this notion of truth. I will argue, here and in the remaining chapters, that we can accommodate a Wittgensteinian pragmatism *and* a realist conception of truth only on the condition that we give up on mentalism and the idea that we possess only a single mode of epistemic access to the world and its objects.

The most influential and best remembered of Frege's contributions to the project of understanding language and language use is his seemingly simple but formally elusive distinction between the "sense" and "reference" of singular terms, and the corresponding distinction between the cognitive content (Thought) and reference (truth-value) of sentences. This distinction, posited to remedy Frege's self-perceived failure in the *Begriffsschrift* (1879) to make his theory of the content of sentences compatible with their supposed truth-compositionality, receives its earliest (and most complete) treatment in *Uber Sinn und Bedeutung* (1891). Nevertheless, because a theory is generally best understood from the perspective of the problems it was posited to solve, we will take a brief look at the *Begriffsschrift* theory and its problems before tackling Frege's later, more difficult work.

The *Begriffsschrift* theory of language holds that the content of a sentence is "that [which] can become a judgement," i.e., that which can be judged true or false.[2] Roughly speaking, it is that which is conveyed by (the meaning of) the sentence. Frege writes: "In the two propositions 'The Greeks defeated the Persians at Plataea' and 'The Persians were defeated by the Greeks at Plataea' . . . [e]ven if one can detect a slight difference in meaning, the agreement outweighs it. Now I call that part of the content that is the same in both the conceptual content."[3] That to which one assents and in which one believes when judging the truth of a given proposition is its conceptual content. Blanchette notes: "Because to judge what a sentence expresses is true is to believe what a sentence expresses, [conceptual] contents are the objects of belief. [Conceptual] contents are also the primary bearers of truth and falsehood; the [conceptual] content, not the sentence, is properly said to have a truth-value."[4]

Now, the principle of compositionality holds that this conceptual content of a sentence is determined by (composed of) the contents of the parts of that sentence, and this relation holds in such a way that terms (parts) of different form but identical content may be freely substituted for one another without altering the content (or the truth-value) of the sentence as a whole. Since for Frege two sentences have the same content if and only if they play the same role in an inferential sequence (hence the intuitive connection between the conceptual content of a sentence and the fact it expresses), substitutions which do not alter this role also leave the conceptual content unchanged.

But, the two roles that conceptual contents are designed to fill—being the object of belief, (that which is understood by a person in judging a sentence true or false), and being the logical content, (that which "influences the possible consequences" of a sentence in a judgement of inference[5])—can be pried apart with clever utilization of the principle of compositionality. Consider the following pair of sentences:

(1) The morning star is bright.
(2) The evening star is bright.

Given that "the evening star" and "the morning star" are co-referential (and since the contents of singular terms are their referents), the two terms are freely substitutable and the two sentences must have the same conceptual content. And, indeed, it seems fairly clear that, insofar as inference is concerned, the two sentences convey the same information and thus have the same "possible consequences" in an inferential sequence. Nevertheless, it seems equally clear that an agent can understand both sentences without knowing that the morning star is the evening star, and thus can believe (judge to be true) sentence (1) while disbelieving sentence (2). If the conceptual content of the sentence is the object of belief, it seems that (1) and (2) must have different conceptual contents. "That is, the belief whose object is expressed by (1) is not the same belief as that whose object is expressed by (2), since one can have the first belief without having the second."[6] Thus, sentences which express the same information, have the same truth conditions and value, and which therefore play the same role in logical inference—whose "logical contents" are the same—may nevertheless express different "belief contents." The *Begriffsschrift* account, which assigned the same entity to expressing "logical content" and to expressing "belief content," thus reveals its inadequacy.

In order to address this failure, Frege proposes the familiar sense/reference distinction (for singular terms) and the corresponding Thought/truth-value distinction for sentences. That part of the content of a singular term which contributes to the "belief content" of a sentence is its "sense," and the part which determines the truth value of that sentence, by being the object of predication, he calls its "referent." Frege maintains that for every singular term there corresponds not just an object designated (the referent), but also a "mode of presentation of the thing designated."[7] This "mode of presentation" is the sense of the singular term, and is roughly the way in which an object is named—or "picked out"— by the term in question.

The corresponding entity for sentences is the Thought, which is the sense of a sentence and is therefore, by the principle of compositionality, composed of/determined by the senses of its parts. Frege writes: "a statement contains (or at least purports to contain) a Thought as its sense; and this Thought is in general true or false, i.e. it has in general a truth-value which must be regarded as the reference of the sentence."[8] Thoughts maintain some of the characteristics associated with the conceptual contents of the *Begriffsschrift*—they are what I have informally designated the "belief content" of the sentence, that which is understood, (dis-) believed, or grasped when judging the truth-value of a sentence. They are also that which *is* either true or false, that which (again like the conceptual content) bears the truth or falsity of the sentence. The difference comes in the way Frege treats the rough equivalent of the "logical content" of the *Begriffsschrift*. The criterion for determining this content is slightly different. Namely, the Thought is that which a sentence shares with all and only those sentences which must be accepted or rejected together, according to a rational agent who understands the sentences in question. Two sentences express the same Thought if and only if an agent who understands both is logically bound to assign them the same truth value. Thus, all sentences which express the same Thought play the same role in an inferential sequence, but not all sentences which play the same role in an inferential sequence (which have the same truth conditions) express the same Thought. It is this split which allows Frege to solve the problem caused by co-referential singular terms.

Of course, from the above it follows that if it is possible for an agent who understands two sentences to judge their truth-values differently, the sentences must express different Thoughts. This is the "criterion of difference" of Thoughts, and is the acid test most often used in Frege scholarship. The criterion of difference indicates that Frege accepted what Evans has designated "Russell's Principle," which states that in order to make a judgement about something, one must know what object one is making a judgement about. In order to judge the truth-value of a sentence, one must know which Thought one is judging—one must grasp the Thought being judged—and this grasp must be sufficiently firm so as to guarantee complete and degree-less access to that Thought. It is not possible to grasp a Thought partially, for in grasping a Thought an agent must be aware of whatever aspects of the Thought are sufficient to differentiate it from all other Thoughts (or the principle of difference would not hold) and these aspects identify the Thought as such; to grasp it in a way sufficient for this differentiation is to grasp the entire Thought: "everything that the sentence expresses which is relevant to the judgement of the truth of what is expressed."[9]

A nice diagrammatic overview of Frege's conception of linguistic structure is provided by Frege himself in his letter to Husserl of 24 May, 1891.[10]

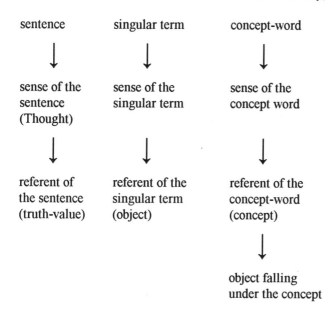

sentence	singular term	concept-word
↓	↓	↓
sense of the sentence (Thought)	sense of the singular term	sense of the concept word
↓	↓	↓
referent of the sentence (truth-value)	referent of the singular term (object)	referent of the concept-word (concept)
		↓
		object falling under the concept

Here the sense of the singular term must play the dual role of providing the completing sense that constructs the Thought by combining with the sense of the predicate, and also of presenting the referent of the term. Likewise, the Thought must both reflect the cognitive content of the sentence (be that which is [dis-] believed or understood when judging the truth of a sentence, and that which is shared by all sentences which a competent language user is bound to accept or reject together) and also bear the referent of the sentence (be that which *is* either true or false—that "for which the question of truth arises"[11]—regardless of anyone's judgement on or grasp of the Thought in question).

The close relationship of sense to truth value also shows up in another way. It is a consequence of the criteria of difference that for an agent to judge two sentences to express the same Thought is for her to judge that they must be assigned the same truth-value in all cases; this is another way to say that she believes two sentences to express the same Thought just in case she believes them to have the same truth conditions. If however, the "truth conditions" of a sentence are those objective factual conditions which must obtain if the sentence is to refer to the True, then we must remind ourselves that while sameness of Thought implies identical truth conditions, the reverse does not hold. Although it is certainly true that sense is related to truth conditions by designating the facts about the world—by definition, relations between concepts (like white) and objects (like snow)—which must obtain for the sentence to be true ("The snow is white" is true if and only if the snow is white, or, in Fregean lingo, if and only if snow falls under the concept white), it is worth emphasizing that sense is not the unadorned display of these referents (in virtue of whose relation the truth-value of the sentence is determined), but rather only one mode of presentation of them. Because of the relation between the sense and referent of a singular term, the objects and concepts whose relation determines the truth-value of the sentence are designated *in a way* by the senses of the parts of the Thought; the Thought itself is a mode of presentation of the truth value of the sentence. Thus, in grasping a Thought I understand (in a way) what must obtain for the sentence to refer to the True. For if knowing "The morning star is bright" was to know objectively the factual conditions which must obtain for the sentence to be true, no agent could fail to see that "The evening star is bright" has these same conditions. We can only conclude that the possibility of our failure to see this indicates that we do not grasp the "truth conditions" of the sentence in understanding it. We must tread carefully: it is a requirement of Frege's semantics that grasping a Thought entails understanding everything relevant to judging its truth value; thus my understanding of the sentence must also be my understanding of what must obtain in order for the sentence to be true. It is tempting here to say that the Thought is itself the truth condition of the sentence, that which must obtain if the sentence is to refer to the True, but we must resist this temptation. For it is only one way of understanding that in virtue of which the sentence would be true. Frege requires that understanding be sufficient to allow the formation of a judgement of the truth of the Thought; we can fulfill this requirement by saying that grasping the Thought is grasping the assertability conditions for the sentence. It may turn out, of course, that we can only account for the competence and accuracy of our truth-judgements by identifying assertability conditions with truth conditions—there is already quite a strong connection given that assertability conditions represent an epistemic perspective on truth conditions. But the notions are not the same, and for precision's sake we should begin at least with the distinction.

Central to Frege's theory of language, then, are the following features: The same entity—the Thought—must both:

(1) "Reflect the cognitive content" of a sentence. This entails that sentences which share cognitive content (which possess the same assertability conditions) must express the same Thought.

(2) "Bear the referent" of a sentence. This entails that sentences with differing truth conditions must express different Thoughts.

Likewise, the sense of a singular term must both:

(1) Provide a completing sense which can join with the sense of the predicate to compose a Thought.

(2) Present one (and only one) particular object.

By definition, the sense of the sentence must be equivalent to the Thought expressed, and, in being constructed from the senses of the sentence-parts, must vary if and only if they vary. If Frege's theory is to be successful, all these features must harmonize into a complete linguistic picture.

2.2: On Sense

Although it is arguable that the Thought is the central and most important aspect of Frege's theory of language, it is the sense of singular terms which we will be concerned with here. Because the sense of a singular term has the dual roles designated above, and most especially because the sense of a singular term, by presenting an object "in a way," must also be a bit of mental content (that which is understood or known about an object) capable of being part of a truth-bearing belief, the sense of a singular term is the ideal point of investigation given our present concern with knowledge of objects.

We noted at the end of the last chapter that, because of the criteria of difference for Thoughts, the sense of a sentence and its assertability conditions must be understood to be intimately related, perhaps identical in an important way. We can begin to understand the implications of this for the sense of the singular term by considering again the two sentences "The morning star is bright" and "The evening star is bright." It is absolutely essential to Frege's philosophy (and, I suspect, to any reasonable semantics) that the capacity to understand sentences entails the capacity to judge their truth-value, and that in grasping a Thought I thereby grasp that which must obtain in order for the sentence to have the semantic value True. Thus, I can only judge one sentence to express a truth the other does not if I take the two sentences to have different truth conditions, and given identical predicates they can only have different truth condition if the names designate different objects. Because my capacity to judge the truth of the sentence must entail a capacity to identify the object described and named therein,[12] I can only attribute my differing judgements with respect to the two sentences to different criteria of identification corresponding to the two names. Thus, even though there is a sense in which the objective criteria of identity must be the same for "morning star" and "evening star," still the criteria of identification (by which I judge their identity) must be different. What all this entails, then, is that the sense of the proper name (singular term) must be (provide) the criteria of identification for the object designated; the mode of presentation must provide epistemic content sufficient for the identification of the object in question, with the understanding that there may be different, equally legitimate modes by which this identification (the differentiation of the designated object from all the things it is not) can be accomplished.

Regardless of how, in the end, it is best to understand such criteria of identification[13] there are certain things which must be common to any understanding. The sense of the singular term must provide a "way of finding" the referent, some method or information by which the referent can be recognized, and also must provide a "way of thinking about" that referent, that perspective by or under which we conceive the referent, and from within which we would make judgements regarding the identity of, and the truth of statements about, said object. Sense must provide the former to allow for our ability to judge the truth-value of sentences containing the name in question, and the latter to account for the possibility of judging differently of object "morning star" and object "evening star": the possibility entails that one possess different criteria by (or under) which one judges the identity of the designated object(s).

2.3: On the Use of Proper Names
    It should be noted immediately—and we will return to this point—that the Fregean account of reference *assumes* that names refer to distinct objects, objects which have objective criteria of identity quite apart from the criteria of identification by which objects can be identified as designated. The only reason that "Ateb" = "Afla" is because there is some definitive object, possessing an identity independently of the criteria associated with either name, in virtue of which both names can refer to the *same* thing. This underlies the fact that sense cannot, in being the agent of differentiation of objects (that by which an object is presented in a way) also be the agent of individuation of objects as such.[14] Objects not sharing the same boundaries are not the same object, and insofar as the senses of the two names are different, and possess different criteria of identification, if their function was not relegated to allowing the differentiation (the *act* of identification) of the object by the agent, and instead also defined the identity of the object (individuated the object as such), the names would refer to *different* objects. Thus, sense must allow discrimination between objects, and the identification of some particular (designated) object, but this ability is rooted in the prior individuation of objects as such (and objective—intersubjective—agreement with respect to this individuation). How this individuation is effected is left an open question by Frege's philosophy.
    This same point can be made to appear in other ways. Let us consider, for a moment, the capacities that must be involved in the proper use of names and other singular terms. To use a name properly is to assign a label to some particular object. But it seems that even more basic than this is the ability to assign something to a kind, and thus to be able to recognize kinds, an ability which may obtain quite independently of one's ability to assign a proper name to some particular example of the kind. Take for example the singular term "the red lamp on my desk." Ignoring for the moment the complications introduced by the context-dependent location and indexicality of the expression, surely identifying the "red lamp" so designated assumes the ability to identify lamps in general, an ability which must be related to one's understanding of the concept-word "lamp."
    Analogously, it does not seem too far-fetched to unpack the sense of a proper

name like "Moses" as, in part, designating "the person Moses," implying an ability to identify persons as a criterion for identifying Moses. It would be very difficult to imagine the capacity to understand and use the proper name "Moses" without the knowledge that Moses was a person. If this is anything close to correct, then we can understand the "way of finding" embodied in the sense of "Moses" as, in part, a way of finding persons, and in part a way of distinguishing the person Moses from other persons. It is, I suppose, conceivable to know that some particular thing bears name "A" without knowing in detail what sort of thing A is. I may know only that "Bubba" of the local zoo is some sort of animal (and maybe what he "looks like"). But it strikes me that as this knowledge of kind gets increasingly scant, so too must our grasp of the name. If I don't know *anything* about Bubba's type classification—animal? vegetable? mineral? gaseous substance? metaphysical construct?—it seems unlikely that I can wield that name to identify any individual. This general intuition seems even stronger if we imagine starting with the name and not the particular: to be capable of using the learned sense of a name to identify its bearer, we need at least to know what sort of thing we are looking for and, more or less, where to look.

Consider, by way of comparison, "the morning star." If the above account is correct, part of the sense of the singular term would enable us to know that the designated object is a heavenly body. But this cannot be taken to imply any sort of correct knowledge of the nature of the object designated. The people who first identified the morning star had very odd ideas about just what heavenly bodies were, and of course it was eventually discovered that the morning star is really the morning planet. In virtue of what, then, have we been able to identify the same object "morning star" for a couple thousand years? If we are to stick with a Fregean theory of reference, our ability is made possible by our possession of a "way of finding," which needn't be more complicated than knowing where to look for the object (the context it is likely to be discovered in, i.e., where to find examples of the kind of thing it is), and how to distinguish the particular "morning star" from other examples of the same kind. These abilities and the information in virtue of which they operate may or may not be understood in terms of descriptions like "the brightest heavenly body besides the sun and moon"; I am inclined to think that such reductions of senses to descriptions are a mistake—after all, the definite descriptions have their own senses which are by definition different from the senses of the proper names they are meant to be related to.[15] But regardless of the details, if the above account is on the right track then it should be possible to examine separately these two aspects of sense—that whereby we identify kinds and how to find them, and that whereby we distinguish a particular example of the kind.

2.31: Direct Realism

Before interrogating these two abilities further, it is worth a brief digression to deal with a certain explanation of our ability to refer which may begin to look attractive while mired in the complexities of Fregean senses: direct realism. I should say at the outset that I have only two modest aims in this section. I hope to show first that some of the important features of Direct Realism can (indeed, in some cases must) be accommodated within a Fregean framework, and second that adopting a Direct Realist model of reference nevertheless leaves one with the same problems and unfinished tasks which, I am arguing, await the Fregean. I take the opportunity of this digression more to get clearer on the requirements for any view of reference than to defend some particular account.

What makes the Fregean picture of reference and object identification so initially attractive is the idea that our ability to refer to objects is intimately related to descriptions; the idea that our ability to refer to things would be related to our knowledge of said things and their characteristics is quite intuitive: names refer because in some sense they describe objects, or are associated with accurate mental depictions of the designated objects.

But the myriad counterexamples which can be generated force a more careful appraisal of this initially intuitive claim. If the ancients thought of stars as holes in the celestial fabric—and nothing exists which meets this description—then they seem to have referred to nothing at all, and we seem to lose that in virtue of which we can refer to the same objects which they (thought they) did. Stars for us, after all, are great flaming balls of gas. The reason we refer to the same objects—and, for the Direct Realist, the reason we can refer at all—is because there is some definitive object which is the same, regardless of our conception of it. So far this is just a restatement of the Fregean thesis that there must be an object, defined independently of our ways of understanding it; to think otherwise is to conflate our ways of *identification* of an object with its criteria of identity. Thus, to understand senses as conceptions of the nature or identity of the object is a mistake from the start; our ability to find or identify an object must be understood in terms of more indefinable "looks" and "modes of presentation" which are intersubjectively and crosstemporally stable and possessed in virtue of grasping the objective sense of a name. In other words, the Fregean can accept this part of the Direct Realist contention, that reference depends on the object possessing criteria of identity independent of our understanding, and nevertheless still insist that our capacity to refer to the object (and to the same object over time) is related to our possession of "ways of finding" and identifying said object.

Other critiques are more difficult to brush aside, as they claim that *no* sort of descriptive knowledge need accompany my references;[16] our ability to refer cannot be related to our possession of some finite set of definite descriptions which uniquely identify the referent, or which we can *use* to find the referent, for the simple and obvious reason that most of us cannot articulate such descriptions for most of the names we use everyday. To account for this, the Direct Realist postulates that reference must be possible in virtue of some connection to the actual natures of objects, regardless of our conceptions of them, and that reference (this connection to the natures of objects) is not related to articulable knowledge at all; names do not have senses, only referents, and our ability to refer depends only on the nature of that referent and not on our possession of any "modes of presentation" or "ways of finding."[17] But it seems to me that the Fregean can quite well accept this critique as well, without thereby accepting the Direct Realist result. She can still insist that we use names properly only in virtue of our possession of some cognitive content sufficient to allow us to find and identify the object.[18] Consider the following:

It seems quite plausible, as Putnam suggests, that one can refer to beech trees even without knowing the distinguishing characteristics of beeches. Philosophers like Putnam and Kripke suggest that we can do so in virtue of our relation to experts who *do* possess knowledge sufficient to distinguish beeches from other sorts of trees. But this seems to beg the question in important ways: if we are to count the experts as possessing the capacities which in fact allow reference, then this looks more like support for a Fregean thesis than an objection; genuine or primary reference occurs in virtue of a capacity for differentiating the named objects, a capacity which obtains just in case one knows the distinguishing characteristics of the object named. I don't see that this knowledge needs to be able to be manifested in the form of an articulable list of definite descriptions in order for this position to count as a kind of Fregeanism; nor need the existence of secondary usages of a word which are parasitic on primary use be antithetical to the spirit of Frege's philosophy.

For the Direct Reference view becomes highly implausible if we want our ability to refer to beeches in this limited secondary sense to imply that we thereby possess the capacity to use the name properly in every situation. Indeed, it seems that one's ability to use "beeches" properly at all is dependent on possessing *some* knowledge of what they are, or a limited capacity to differentiate them from that which they are not (I take it that these two things are generally synonymous): we must know in the first instance that they are trees, and increasingly sophisticated uses of the term are likely to be possible only on the condition of acquiring increasingly sophisticated knowledge or capacities for differentiation. Fregean accounts associate these capacities with the semantic value of the name, that which one grasps just in case one understands the name and can use it properly. Insofar as this entails simply that one acquires the proper use of a name (understands the name[19]) just in case one has the relevant capacities for identifying its bearer, then something like this is surely correct.

The overall intuition operating here, nicely put by Howard Wettstein, is that "in order to be thinking about something one must have a cognitive fix on it, . . . something in one's thought must correctly distinguish the referent from everything else in the universe."[20] What is distinctive about Frege's claim is that this "cognitive fix" is entirely cognitive, and intimately related to linguistic understanding; our capacity to differentiate objects rests on the possession of the proper mental content, the grasp of the objective sense of a name. As noted, the Direct Realist is content with less stringent requirements, often involving certain special causal connections to the object of reference, even causal connections mediated by other persons.

But however, in the end, these positions are to be understood, and however the differences between them are settled, they can at most apply to the differentiation of already individuated objects from one another—which, as I have been insisting, leaves a whole area of inquiry wide open.[21] I do not expect the above to provide any definitive reasons for accepting one position over the other (although I do hope to show that Direct Realism is not *obviously* more attractive than a Fregean account); whatever the merits of Direct Realism,[22] my main intention here is to point out that it does not escape the very general critique I am making against Fregean epistemology: that there is a deferred project of understanding how we come to know objects as such, how it is possible to individuate objects. The Direct Realist offers another model of that in virtue of which we differentiate objects from one another, and refer to these differentiated particulars. But it has not, cannot have, provided us with a way to understand the individuation of objects. For if we are to make sense of an ability to refer to objects independently of our conceptions of them, then it can only be in virtue of the fact that objects are already presented to us as such; objects cannot be the contents of proper names if the names are that in virtue of which the objects are individuated (for this would indicate content apart from—prior to—the object).[23] Unless we imbue language with occult powers which it may possess independently of our use and understanding of it, then we must agree that all there is to language and its meaning is what we grasp in understanding it; names are meant to designate objects, which implies the existence and individuation of objects to be named. However we understand and explain the presence (individuation) of objects, then, Direct Realism does not contribute to this level of understanding.

2.32 The Identification of Kinds

Although it is almost certainly not true that every identification of a particular presupposes the identification of a kind more explicit than "object," it is highly likely that every identification of a particular in virtue of a name (or other sense bearing singular term) *does* presuppose the prior differentiation of a kind, the narrowing of the search to entities of a particular sort. If the sense is to provide a way of finding its referent, understanding the contexts in which the search will take place must be a primary contribution.[24]

As I noted already, Frege's distinctive claim for sense is that it explained that in virtue of which we are able to refer to and identify objects in terms of our possession of a certain cognitive content, which content Frege associated with the grasp of the sense of a name (the same grasp which allows the proper use and linguistic understanding of that name). The critiques of such figures as Saul Kripke have taught us that whatever this cognitive content associated with the grasp of sense amounts to, it need not (generally does not) manifest itself via the ability to articulate a set of uniquely referring definite descriptions for each name one understands. For insight into a *Fregean* account of our capacities vis-à-vis names and kinds, then, it will be wise to look to those distinctively cognitive capacities which must obtain for the proper use of words in the language, and not be too quick to associate these capacities with the possession of articulable object-descriptions.

Still, we can hardly get away from the fact that our ability to identify objects and kinds (and objects *as* examples of kinds) must be related to the epistemic/perceptual availability of occurrent properties of said objects and kinds. We have no hope of discerning a kind unless we perceive an example of it, and no differentiation is possible if this perception fails to make available information about the object sufficient to effect the necessary identification. Thus, whatever cognitive content is supplied by grasping the sense of (and properly using) a name, it can only be useful for an identification if it is somehow related to epistemically available information about the object (or object-type) in question.

We can see why perception-related definite descriptions were such an attractive option for those wishing to explain object differentiations within a Fregean framework. But there are no doubt meaningful things we can say without recourse to these obviously problematic entities.[25] We might start by stressing, in the words of Dummett, that

> the sense of an expression constitutes the manner in which something is determined as being its referent, or, to put it in terms of our understanding of the expression, that is, of what is involved in grasping its sense, to grasp the sense is to apprehend the condition for something to be its referent.[26]

It is evident that this interpretation of sense must apply not merely to sentences (we noted earlier that grasping the Thought is grasping the conditions under which it refers to The True) and proper names (whose conditions of identification are so given) but also to predicates, insofar as they contribute to the truth conditions of the sentence by both providing a completing sense for the expression, and designating the concept under which the named object must fall if the sentence is to be true. To grasp the sense of the predicate, then, is to grasp that which would have to obtain for the object to so fall, or, considered apart from its place in the whole Thought, to grasp the conditions to be met for any object to fall under the designated concept. Turning again to Dummett:

> the sense of the proper name determines an object as its referent, and the sense of the predicate determines a mapping from objects to truth-values, that is to say, a concept; the sentence is true or false according as the object does or does not

fall under the concept, that is, according as it is mapped by it on to the value True or the value False. The mapping of objects on to truth-values is not the sense of the predicate, but its referent: the sense is, rather, some particular way, which we can grasp, of determining such a mapping.[27]

What we have come to so far is that grasping the sense of such predicates entails grasping the conditions for falling under the concept thereby designated, and that whatever these conditions are, if they are to be practically useful, they must be related to epistemically available information about the objects in question, information about the properties of the objects in virtue of which they would fall under a given concept.[28]

We should turn, then, to an examination of those epistemic capacities in virtue of which we glean the relevant information about the objects of our world; the nature of these capacities and of the information they provide will restrict the range of allowable conceptions of sense.

I hope it is clear by now that for Frege, that in virtue of which we differentiate and identify objects is cognitive in nature, a mental content associated with the act of linguistic understanding—grasping the senses of names. Frege also believed in only one mode of epistemic access to the world, and I hope to show that insofar as this is true, that his understanding of our discrimination of objects and classes must be parasitic on prior individuation of these entities by other means.

So what did Frege think about the character of our epistemic access to the world? A passage from *The Thought* should prove especially illuminating:

Sense perception indeed is often thought to be the most certain, even the sole, source of knowledge about everything that does not belong to the inner world. But with what right? . . . Having visual impressions is certainly necessary for seeing things, but not sufficient. What must still be added is not anything sensible. And yet this is just what opens up the external world for us; for without this non-sensible something everyone would remain shut up in his inner world.[29]

There is much to be said about this passage. First of all it does not challenge the idea that Frege held there to be only one mode of epistemic access to the world; what this questions is whether this mode of access is restricted to perception. For if we can only think of objects in one particular way (at a time), then this implies that all our modes of presentation (all our ways of differentiating and identifying objects) are on the same epistemic level and therefore mutually exclusive, else we could be thinking about something in two ways at once—fixing the referent by two modes of access. The idea that our access to the world is aspectival in this way, which is very basic to Frege's thought, rests on the notion that we have a single mode of epistemic access (a single informational conduit, if you will) to the external world.

Next it should be noted that this passage should have given pause to the myriad commentators and interpreters of Frege who immediately assimilated senses to sense-perceptions and sensation-based descriptions. Frege clearly indicates that senses are meant to include something "non-sensible" which participates in the individuation of objects. Thus this is evidence for a non-sensible cognitive capacity for the differentiation of objects and classes.[30] None of this challenges what we have said so far, and if we were to say quite simply that sense is whatever information we use to differentiate objects, then clearly something must fit this description, since we do, in fact, succeed in our object and class differentiations. But there are important restrictions on sense which must continue to be observed: senses are cognitive contents related to linguistic understanding, whose criteria of identity are intimately related to our judgements of the truth value of sentences (their identities, that is, can be understood in terms of what I have called the assertability conditions of sentences).

Thus it is quite wrong to conclude, as Dummett does, that

[Unlike for] Mill, [who] wrote as though the world already came to us sliced up in objects, and all we have to learn is which label to tie onto which object . . . [for Frege] the proper names which we use, and the corresponding sortal terms, determine the principles whereby the slicing up is to be effected, principles which are acquired with the acquisition of the uses of these words.[31]

Given the relation between sense and truth conditions, there could be no such thing as truth-preserving substitutions of different, co-referential names if the names carried with them senses which determined the criteria of *identity* for objects. For insofar as the criteria of identification were different, the objects would be sliced up by different principles, and thus fail to be identical; without the guarantee of identity there is no true co-referentiality, and so it could only be accidental when the truth-value of sentences remained the same for such substitutions. Nor can we say that the criteria of identity which are grasped are the same for differently named identical objects without the senses of the different names also being identical. We are forced to maintain, as I have throughout this chapter, that the sense provides criteria of identification which differently point to objects with the same criteria of identity. That by which truth is determined (the actual identities and properties of objects) cannot be conflated with that on the basis of which truth is judged without lapsing into a kind of idealism whereby all truths are analytic—as if all we judge to be true is true because our judgements of truth determine the identities of objects.[32] We could not make sense of the possibility of incorrect judgements without disconnecting the practice of judgement from the definition of truth (and interpreting mistakes in terms of straying from community agreement will not suffice: we must be able to understand entire communities or historical eras as being mistaken); yet we cannot make sense of our epistemic success without supposing we have access to that which defines and determines truth. Needless to say, this is a very delicate area of philosophy which calls for treading lightly, but a few remarks are in order.

There are two issues here which need to be explicitly separated: we must be able to make sense, for Frege, of different terms referring to the same object, and we cannot if we charge sense with the function of ontologization, of individuating objects (or providing the criteria of their individuation). Likewise, the Thought refers to a particular truth-value in virtue of the fact that its parts designate objects and concepts, whose relation may or may not obtain as stated (Frege is quite firm that this occurs quite apart from anyone's grasp of the Thought). Let us suppose that no one knew that Bruce Wayne was Batman (perhaps neither Bruce nor Batman knows, because of some special mental effects associated with the transformation): we can suppose perfect competence in using the names, a full grasp of the sense of each name, without the knowledge that they are the same being. However, we can only understand the possibility that they are the same being, the truth of the statement "Bruce = Batman," on the condition that the senses are only criteria of identification, and not also criteria of identity. Even statements like "Batman lives in Wayne Manor" would be false on Dummett's supposition that senses "slice up the world,"[33] and any statements we could make about either which involved essential attributes would be true not just necessarily, but analytically by linguistic definition.[34] I take it that these are undesirable consequences, and they underline the fact that a reasonable notion of truth requires a bit more distance between language and reality than Dummett allows.

We have not gone as far afield as it might appear. We noted earlier that to grasp the sense of a sortal predicate was to grasp the conditions under which an object would fall under the concept designated. But we face the same dilemma here negotiating between assertability conditions and truth conditions as we did above between criteria of identification and criteria of identity.

We saw that, even on Dummett's own interpretation of predicates (which I think is correct) their reference is the function mapping arguments onto the value True; "X is a human male" maps a given set of arguments onto the value True. What makes it true for any particular argument is that it gets mapped onto the value True. We must note, however, that "X is a human with a Y chromosome" refers to the same function mapping certain arguments onto the value True; it is true for the same set of things and that in virtue of which the truth is determined (the function) is the same for both predicates.

Thus, we must understand the sense of the predicate to be a certain epistemic perspective on the referent of the predicate, i.e., the function mapping arguments onto the value True. The predicates have different senses because they represent different ways of understanding the same mapping.[35] Grasping the sense of the predicate then, is grasping in a particular way that in virtue of which any given function would fall under a sortal predicate. Thus here again we must insist that the information we glean in grasping the sense of a predicate are the conditions under which we would judge an object to fall under the concept named. Sticking with the requirement which makes grasping the sense epistemically useful, i.e., that the content grasped be related to epistemically available information about the object in question, we can say that the sense of a predicate gives a perspective on those characteristics in virtue of which an object would be mapped by the named function onto the value True.[36] These characteristics must be occurrent (although not necessarily sensible nor propositionally articulable) properties of objects, and might include such information as where to look for such a kind, what sort of space/environment it might occupy,[37] its function, and whatever other information one might need to exercise a capacity for discriminating between kinds. We should always bear in mind, too, that this same capacity is expressed in the proper use of given predicates; the capacities we exercise to distinguish males from females, for instance, are quite different from the ones which a scientist might use to determine who possesses a Y chromosome.

Thus, in answer to our very general question as to how we divide objects by kinds Frege has this very general answer: we are able to divide objects by kind because we have an epistemic perspective on that in virtue of which they are in fact so divided. We acquire this perspective, and grasp the cognitive content which makes it possible, in coming to learn the proper use of the language, and of a particular name or predicate within that language. This alone should be enough to show that the perspective we acquire in learning the use of the term is not to be conflated with that in virtue of which objects fall under sortal concepts; if the use of a term was that in virtue of which its referent was defined, then having the use of the term would not be acquiring a *perspective* on the agent of differentiation, but would be conceiving the bare agent of differentiation itself, the "sortal nature" of the class. It is clear that Frege thought we acquired no such thing, and a fairly safe position that he was right to think so.

However, in taking this position, one is immediately faced with a dilemma. As soon as we admit that meanings are expressions of our limited epistemic perspectives on "objective" reality we open the possibility of scheme-content divisions: we may not be able to know the actual relations of assertability conditions (which we grasp in understanding the predicate) and truth conditions (on which our assertability conditions are merely a perspective).[38] The truth-values of our assertions are potentially beyond verification. The looming spectre of skepticism is poised to make an entrance here, and it is this entrance which Dummett seeks to block by insisting on the conflation of criteria of identification and identity, and, ultimately, of assertability conditions and truth conditions (insofar as they turn on matters of ontologization.) This effort, however well meant, is not only out of sync with Frege's writings, but carries with it a high philosophical price.

Sense is meant to be that cognitive content whereby we differentiate and identify objects, in a way intimately related to proper language use. But insofar as the epistemic access represented by sense is understood to be an aspectival and limited perspective on an already defined reality, we face the potential of skepticism; more practically we cannot account for the apparent accuracy of our judgements (and at the same time head off the skeptic) unless we suppose that we have epistemic access to the structures and states of the world by which truth is determined. The solution cannot be, however, to extend the epistemic *access* to the world represented by and preserved in sense and language acquisition by making language itself define reality; to do this entails jettisoning the aboutness of language, and with it the notion that truth is a measure of the appropriateness of that relation of aboutness.

I will set out these conflicting demands, and the sort of solution they point towards, in section 2.4.

2.33 *The Identification of Particulars*

I have insisted throughout the first sections that the fundamental characteristic of language, a characteristic we want to preserve, is its aboutness. But what, we might ask, echoing Wittgenstein's own query about thought, makes language about anything? To a large degree the burden of this question falls to the reference of singular terms: what makes a word refer to an object? What, in other words, is it that needs to be preserved in order to preserve the aboutness of language?

We have seen that for Frege, what makes a thought about an object is that the agent possess a "cognitive fix" on the object in question, and this cognitive fix consists, at least in part, in possessing information and relevant capacities sufficient to pick out the object in question, were it to present itself. The sense of a singular term provides not only that which allows the agent to pick out the relevant object, but this grasp of sense must also allow for truth-judgments of sentences containing the designating term. These capacities are, of course, intimately and necessarily related; for one has the ability to pick out and identify an object precisely by having an epistemic perspective on that in virtue of which the object is defined. This epistemic perspective provides what I have called the criteria of identification for the object, that which the object must meet for an agent to judge something to be the object named, and, of course, possession of these criteria are meant to be sufficient to allow judgements regarding the truth of statements which refer to said object.

It is obvious, given the many-one relation of senses to referent, that sense, in providing different criteria of identification for the referent, cannot likewise provide the criteria of identity for it. But this reveals something important about reference: whatever information regarding the object to be identified is provided by sense, it cannot be sufficient to define the boundaries of that object for the agent, and so can only allow reference to the named object on the model of deferred ostension—"the object which conforms to such-and-such." Yet if the agent is to experience the named object as a determinate particular then its boundaries must nevertheless be present to the agent; to know that it is *this* object which is meant is to know it *as* an independent, well-defined object. For this to be possible the agent would need some form of epistemic access to that in virtue of which objects are individuated as particulars. But sense is specifically prohibited from giving this level of information; to conflate the criteria of identification for an object with its criteria of identity would be to lose that in virtue of which different terms could be co-referential, and insofar as sense would be meant, on this conflation, to contribute to our experience of the object (the modes of presentation and ways of differentiating whereby we come to know the object as such), we would lose that in virtue of which we could know or experience the referents of supposedly co-referential terms *as* identical.

Still, if we deny the agent possession of information according to which objects could be individuated as such, we cannot make sense of the possibility of reference to determinate particulars, for whatever the *discriminating* information provided by sense, without *individuating* information the agent would lose that in virtue of which she could distinguish between the identification of a part of an object meeting the discriminating criteria, the identification of the object meeting the criteria, and the identification of several objects all meeting the criteria of identification. In order to be able to know and recognize an object in virtue of given information (which information must represent epistemically available occurrent properties of the object in question) then an agent must already possess the capacity to divide the world into objects, or she would never know which observable characteristics belonged to a particular object; she would not be able to differentiate between observing the defining edges of an *object* and observing the defining edges of a part of an object possessing the looked-for characteristic. And if she does not possess the capacity to make this differentiation then she surely cannot be said to possess the capacity for picking out a determinate object, which capacity is meant to be an entailment of grasping the sense of a term.[39]

For reference to be possible we must presume the capacity to individuate objects as such; the dilemma is that the epistemic access to, and information about, the world in virtue of which this capacity is possible cannot be provided by sense. We must posit a mode of epistemic access to the world, and in particular to that in virtue of which objects are defined, other than the mode of access (epistemic perspective) provided by linguistic understanding.

What is right about Dummett's position on these matters is his distinction between Mill's position and Frege's: Frege is not a simple nominalist, and Mill's apparent nominalism is closer to Direct Realism than to Frege's view. What is distinctive and, I think, right about Frege's position is that learning the name of an object is acquiring the capacity to identify, to pick out, the object named, which capacity Frege associates with grasping the semantic value of the name. This grasp is precisely what is acquired in learning thoroughly the use of the word, a bit of knowledge/cognitive content necessary for both the proper use of the language and the requisite identification of named particulars.[40]

Still, it is worth stressing the central contention of this subsection: our ability to discriminate between objects, and to refer to named particulars, is different from, and depends upon, the capacity to individuate objects and experience them as such. This capacity to know objects suggests that, in addition to the mode of epistemic access associated with our manipulation of linguistic concepts and their operation in sense-perception, that epistemic perspective on the criteria of identity of particulars and kinds entailed by our grasp of sense, we possess another mode of access to that in virtue of which objects are defined and individuated. This mode of access, by the possession of which we come to know objects as such, albeit necessary to the possibility of reference, was left unexplored and perhaps even unrecognized by Frege. This is the deferred project of understanding individuation which careful interpretation of Frege's writings reveals, and it is this deferred project which the bulk of this work is meant to address.

Before moving on to the last section, however, it will be useful to present another claim, a claim for which I have little evidence at this stage, but which I hope is in itself plausible: the distance between criteria of identification and criteria of identity which is central to Frege's theory of reference is a necessary condition for reference itself. To refer to a thing is to name, point toward or indicate a thing, a thing already defined in abstraction from one's reference. I do not think one can make sense of the intentionality of the act of reference, of the "pointing at" involved, unless we take the thing indicated to be separate from (to possess a degree of autonomy or independence from) the act of ostention. In this sense, I take the distinction between criteria of identification and identity (expressed by Frege as the distinction between sense and referent) to be necessary for the aboutness that is essential to language.

2.4: The Deferred Project of Individuation

We have seen that, for Frege, we have an ability to distinguish objects by type, and to pick out particulars, in virtue of having an epistemic perspective on that whereby objects are sorted and identified as such. This perspective is achieved in the process of understanding and acquiring the proper use of terms, and thereby grasping their sense. But we have also seen that there are pressures to posit a more direct, or alternate, form of access to that in virtue of which objects are sorted and identified: we want to avoid the possibility of scheme/content divisions for sortal types and allow verification for our predications.[41] Furthermore, access to the identifying boundaries of objects, which cannot be accounted for in terms of sense, appears to be presupposed by the possibility of definite reference, and thus of the possibility of thought's bearing on material particulars. It is common to accommodate this access by positing a kind of linguistic idealism, whereby our apparent knowledge of the boundaries of classes and individuals is understood to result from the fact that the use of a term actually defines the range of its appropriate application, thereby defining the boundaries of its referent. But we have seen that, for Frege at least, this option is not viable; I would like, in this section, to examine the pressures, some of them aspects of Frege's own thought, which make the idealist position attractive. But I would also like, without having to argue against idealism *per se*, to suggest that there are some things worth preserving in Frege's realism, and to offer a direction which might be taken which accommodates both sides of the dilemma.

Negotiating between idealism and realism is a philosophical activity with a rich history and, one might add, an active present. The dilemma is simple: how can we balance the desire that our ideas, impressions and knowledge of the world actually reveal something about the world, accurately depict the world as it is, i.e., the desire that our knowledge of the world is *true*, with the demand that our concepts be humanly usable, thus to an important degree artificial? This dilemma has myriad forms, and shows up in Frege's work as the question of how to balance between the need for both objective criteria of identity and for epistemically useful criteria of identification. To insist on both (as I have been arguing Frege does) leaves open the question of their relation, and here the dilemma arises once again: for truth demands that there be a world according to which truth is defined, and yet if truth is to be a meaningful measure of the accuracy of our statements then we need access to that world.

This access, of course, is meant to be provided by sense: but this is the very root of the problem, for it is precisely the accuracy of this mode of access which is thrown into question by skeptical appraisal. Insofar as it is artificial and geared to human needs and conceptual abilities, sense looks likely to fall short of the world,[42] and yet to extend it to *define* the world's structure is just to close off the very access to actuality which one wants if our thoughts are to "reach" the real.

There are very good and convincing arguments both for wanting there to be a self-standing world apart from our conceptualization, to which we have (conceptual) access, and for the desire that our knowledge be humanly verifiable and the contents of our concepts humanly interpretable, but I would like to suggest that the apparently intractable dilemma thereby created is to a large degree a product of the unrecognized and unquestioned assumption that we possess only one mode of epistemic access to the world, and that therefore both these desiderata must be accounted for by that mode alone.

We ended the last section with the realization that the possibility of reference demanded both epistemically useful information about the objects of the world (a capacity to differentiate objects from one another), and a capacity to individuate objects as such. To force the same mode of access to fulfill both these functions[43] is to endorse a linguistic idealism (the costs of which we have seen), and yet to insist on realism with only one mode of access is at best to leave the question of our individuating capacity (our ability to know objects as such) unexamined, and at worst to invite skepticism.

My intent is to avoid this dilemma by positing the active body as a mode of epistemic access to the world, access which accounts for our capacity to experience objects as individuals and ensures the "humanness" of our concepts and the verifiability of our knowledge, and yet allows sufficient distance between our conceptual perspective on the world and the world itself to maintain the desired aboutness of language and the intentionality it signifies.

We will return explicitly to questions of individuation and boundaries in chapter 4. For the next part I should like to continue to explore the conditions for "aboutness," and for the preservation of truth-claims. I will argue that we can account for thought's bearing on reality (our capacity to think about the world and its parts) only if we come to accept (1) experience is always (already) conceptual, and (2) the world can act as a guide for experience, limiting its contents. As we shall see, to deny (1) is to invite skepticism, while to deny (2) leads directly to an unacceptable idealism. I will be arguing over the course of the next chapters, and especially in the last sections of chapter 4, that we can accept (1) and (2) if we see that the concepts which structure and give significance to experience are themselves open to the world through the active body.

Notes

1.   This section is deeply indebted to Patricia Blanchette's graduate seminar on Frege, and to her article "Fregean Thoughts and Indexicals," which contains an exceptionally clear exposition of Frege's work. I also benefitted from comments made by Carol Rovane on a very early draft of this section. Some of this material appeared originally in "Making Sense of Indexicals."
2.   Gottlob Frege, *Begriffsschrift* in *Frege and Godel: Two Fundamental Texts in Mathematical Logic*, Jean van Heijnoort, ed. (Cambridge, MA: Harvard University Press, 1970), 12.
3.   Frege, *Begriffsschrift*, 12.
4.   Patricia Blanchette, "Fregean Thoughts and Indexicals," *CSLI Report* no. 88-134 (November 1988): 2.
5.   Frege, *Begriffsschrift*, 12.
6.   Blanchette, 3.
7.   Gottlob Frege, "On Sense and Meaning," in *The Philosophy of Language* A. P. Martinich, ed. (Oxford: Oxford University Press, 1985), 200.
8.   Gottlob Frege, "On Function and Concept," in *Posthumous Writings* (Chicago: University of Chicago Press, 1980), 31.
9.   Blanchette, 5.
10.   Gottlob Frege, Letter to Husserl of 24 May, 1891, in *Philosophical and Mathematical Correspondence* (Chicago: University of Chicago Press, 1980).
11.   Gottlob Frege, "The Thought," in *Posthumous Writings* (Chicago: University of Chicago Press, 1980), 20.

12. And we can begin to see already the sorts of epistemic capacities which must be possessed by Frege's linguistic agents—for the ability to identify objects entails not merely that one possess information that could lead *someone* to identify the object, but that the agent herself possesses the capacities to utilize the information (which, as we will see, must be provided by sense) in ways sufficient to identify the object and to judge the truth of some description of that object. (Although, of course, correct judgement cannot be guaranteed, still the epistemic connection between judge and object of judgement must be sufficiently firm so that even an incorrect judgement can confidently be said to have been a mistaken judgement about the application of the description to *this* particular object; it should not normally be possible to redescribe the judgement as a correct judgement about some other object.)

13. I deal at some length with the various possibilities for understanding the senses of singular terms—whether as "ways of finding" the referent, or "ways of thinking about" it—in "Making Sense of Indexicals." But regardless of where one comes down in these disputes, there are very large areas of agreement about the roles and tasks of sense, and I hope to make my points by staying within these areas.

14. I will be using these terms, "differentiation" and "individuation" consistently throughout. Differentiation will refer to the activity or capacity to identify and pick out particular objects as different in a field of already given objects. Individuation refers to the act or ability of distinguishing objects as such, of ontologizing.

It is worth noting that the concept of an "object" is foundational for Frege's philosophy, part of his division of the world into concept and object; it was not his intention to interrogate its meaning, the criteria of its identity. It should not, therefore, be surprising that his philosophy, in assuming the existence and epistemic availability of objects, stops short of explaining their availability as such.

15. Of course, such reductions generally accompany a denial that proper names have a sense of their own. In Searle's "Proper Names," for instance, the sense of a proper name is a function of its associated cluster of definite descriptions.

16. See, for instance Donellan "Proper Names and Identifying Descriptions," Saul Kripke *Naming and Necessity*, Hilary Putnam "Is Semantics Possible," "Explanation and Reference" and "The Meaning of 'Meaning,'" and David Kaplan "Demonstratives."

17. David Braun, in "Empty Names" introduces the following as the "Fundamental Thesis of Direct Reference: A proper name has no semantic function other than referring to an individual" *Nous* XXXII, no. 4 (December 1993): 449-50. He goes on to note that many theories of direct reference make claims that go beyond the fundamental thesis, among the more common and accepted of which are: "The semantic value (or content, or meaning) of a proper name, if it has any, is the individual to which it refers. A sentence (or utterance of a sentence) that contains a proper name expresses a singular proposition, when it expresses any proposition at all." Braun, 450.

Hilary Kornblith, in "Referring to Artifacts" provides an illuminating example of the Direct Reference position:

"On Putnam's account, it is useful to introduce the following philosophical fiction. Speakers introduced natural kind terms into the language by pointing to a sample of some natural kind and saying, 'Let us call whatever has the same nature as this sample such-and-such.'. . . natural kind terms function as if they were introduced in this way. As a result, natural kind terms are not synonymous with any description of the kind in observable terms." *The Philosophical Review* LXXXIX, no. 1 (January 1980): 111.

It might be noted that in such naming ceremonies the individuation of the object prior to its naming *and* its differentiation by the namer as a particular representative of a kind, are both explicit assumptions.

18. I see no reason to balk at calling this content, insofar as possessing it is a necessary condition for proper use of the name, the semantic value of the name, that which is grasped in understanding it.

19. Which is quite different from understanding *that* some linguistic entity is a name.

20. Howard Wettstein, "Cognitive Significance Without Cognitive Content," *Mind* 97, no. 385 (January 1988): 2.

21. I am hoping that my extension of what counts as an epistemic capacity to include the body and its activity will help provide the framework within which this neglected question of object individuation can be answered. It is worth noting, too, that the vast majority of attempts to understand this "cognitive fix" in a Fregean way have focussed on mental representations, sensations and other descriptions, for each of which attempts there have been produced numerous counterexamples. I intend my inclusion of the body and its activity as an epistemic capacity to supplement, too, the still open question of object differentiation: what makes our thought about an object? I think that including behavioral attitudes as *constituents* (and not merely derivative effects) of our object individuations can go a long way to helping understand what the capacity to differentiate and identify objects which must accompany the proper use of names could possibly *be* (if not the ability to articulate applicable definite descriptions). Our agency does not only rest on the individuation of objects, but contributes to it.

22. And in some instances, as for instance John McDowell's *Mind and World*, the merits are considerable. I most certainly do not mean to imply that there are not good reasons for the Direct Realist position, although I do insist that while I agree that their critiques of Fregean accounts are telling, they are not sufficient by themselves to motivate or ground Direct Realism as the only viable alternative.

23. Nor can the name possess content sufficient for that individuation (this is in any case explicitly denied by the Direct Realist), since that would imply that understanding the name would provide insight into the criteria of identity (nature) of the object, and it is precisely our general lack of such insight which motivated the direct realist in the first place.

24. It is, no doubt, possible to claim that senses contain information only for the differentiation and identification of particulars. But the effect of this relegation would be only to point out the existence of yet another capacity necessary for the utilization and understanding of language and its sense, which was nevertheless left unexplored by Frege. Besides, there are the senses of concept and class terms to be understood, and this section is meant to apply to those as well.

25. Besides the problems commonly pointed out in the literature, definite descriptions seem especially unlikely to represent adequately our differentiations between kinds since, according to an objection that is at least as old as Plato, no set of definite descriptions can ever match the object-kind differentiations we actually make. Plato's solution, of course, was to posit that our recollection of the class-unifying Form gave us the epistemic capacities in virtue of which we differentiated and classified objects of the physical and abstract realms. Insofar as Frege is a Platonist, we should expect Senses to be more like Forms and less like descriptions, although we should not expect Frege's philosophy to support anything like Cognitive Intuition.

26. Michael Dummett, *The Interpretation of Frege's Philosophy* (Cambridge, MA: Harvard University Press, 1981), 249.

27. Dummett, *The Interpretation of Frege's Philosophy*, 252.

28. I hope the reasons for focusing on the sense of predicates are evident. If we are to understand the "way of finding" (and, to some degree no doubt, the "mode of presentation") of proper names and singular terms, and if I am correct that the capacity to identify designated particulars depends on an ability to identify kinds, then we should try to understand our identification of kinds—and kinds, like "person," "man," and "tree," are all predicates; they are concepts under which certain classes of individuals fall.

29. Gottlob Frege, "The Thought," in *Posthumous Writings* (Chicago: University of Chicago Press, 1980), 20.

30. It may be possible to understand this, rather than as evidence for some other kind of information to which Frege believed we had access, as a denial that sensible information possesses intrinsic significance apart from shared, objective criteria for determining this significance, criteria which cancel out whatever subjective element there might be to sensation and ensure that separate sensations—subjective information about the same world—nevertheless signify the same thing. This, of course, was the point of Wittgenstein's emphasis on shared practices and "forms of life." If my aim here were to understand Frege's intentions in this passage, I would take this route, although the shared criteria for Frege are more likely to be logical than pragmatic.

31. Dummett, *Frege: Philosophy of Language*, 179.

32. It is too easy to confuse the point I am making with support for a kind of strong realism against a kind of epistemological anthropocentrism. The point is a more general one about the conditions for a working assertion-related notion of truth: as a measure of the appropriateness of statements to their objects (subjects) truth loses its sense if we *define* assertions as appropriate by connecting language use too deeply with ontologization, by making linguistic meaning the agent of individuation. On the other side of the coin, truth becomes incomprehensible (even as a regulative ideal) if we imagine the "reality" by which assertions are to be measured to be humanly inaccessible (what Dummett calls "verification transcendent"). Reality and its significance must be accessible to human conception if any assessment of its relation to linguistic assertions is to be made. It is my intent to use the activity of the body to help thread this needle: to put it baldly, bodily activity is an epistemic agent of individuation which satisfies our need for a degree of epistemological anthropocentricity without sacrificing the distance between linguistic assertions and their objects necessary for truth.

33. Not because, according to this brand of idealism, the mind determines not just matters of ontology but also matters of fact across the board; but some matters of fact *will* be determined, and other possible options will be extremely limited by the determination of an ontology. A matter of fact like who lives in Wayne Manor is one such—only Bruce Wayne and his staff live there, unless it is also true that Bruce = Batman.

34. All this, of course, on the supposition that Batman and Bruce Wayne were not just comic book characters: there is a sense in which all statements about him/them are false given their fictional nature. But Dummett's hypothesis raises the question of whether there could be made such a sharp distinction between the identities of such fictional characters and the identities of real objects as "sliced up" entities. In Detective Comics all the statements made about Batman are necessarily true of Batman, just because they define his identity; to what degree are uses of a proper name to define the identity of its bearer? We

cannot say that all true (proper) uses of a name define its bearer (although Frege might have said something like this) without presupposing a notion of truth/proper use which itself involves the prior definition of the identity of the bearer. One expects that proper uses of a name and the identity of its referent to be correlated, but the cause of this correlation cannot be that use defines identity.

35. We must resist the temptation to talk in terms of sets here. The two predicates have the same truth conditions not in virtue of the fact that they are true of the same set of arguments, but because they refer to the same function which maps members of the set onto the value True in virtue of the same facts. The predicates "has a heart" and "has a kidney" are true for the same sets of things, but they do not have the same truth conditions. For Frege this is because the two predicates refer to different functions whereby the same arguments are mapped onto the True. Because they therefore have different truth conditions they must also have different sense. We must always remember that propositions may have the same truth conditions but different sense.

It would not be wrong, although it would be potentially misleading, to say that the truth condition for any sentence containing the predicate might be put: "being a member of the set for which the referent of 'X is a human male' maps each argument onto the value True." This way of putting things may be useful later.

36. This is, of course, different from saying that the sense of a predicate represents an epistemic perspective on that in virtue of which the function maps an argument onto the value True, for this states that we have a perspective on some characteristic of the *function*. I maintain, however, that if it is to be an epistemically useful perspective, adequate for allowing judgements of truth in virtue understanding alone, then this understanding must entail, too, a perspective on the characteristics of objects in virtue of which they are sorted by the predicate. For surely the characteristic of the function on which one has a perspective is just that characteristic by which objects are appropriately sorted according to *their* characteristics. If this attribution is a failure of Frege scholarship, then it is only so in virtue of a flaw in Frege's epistemology.

37. For a morbid but telling example, we are not likely to look for a person at the bottom of a river, yet we often look for dead bodies there; different sorts of things (even when as closely related as the above) indicate quite different ways of finding.

38. We can begin to see the problem, in so far as it is not obvious, by working through some of the complexities of the example we have been using. It certainly seems possible that our discriminations might not lead us to differentiate accurately (members of) the natural kind "male." Likewise, when we consider that our reasons for believing that possession of a Y chromosome defines the natural kind is really based on the apparent co-extensionality of our set "male" and the scientific set "possessing a Y chromosome." (The latter is often confused with a more accurate way of defining the former, but this cannot be right: whether it is more or less accurate is a determination that could be made only with reference to some third set, considered the "objective" standard. It is right to say, as Frege's analysis makes clear, that they are different ways of determining the same thing.) We realize that it may be that this way of discriminating does not accurately correspond with the objective differentiation. And it is certainly possible that our ways of determining who possesses a Y chromosome are less than perfectly accurate. The problem arises, of course, because in the absence of non-perspectival access to that in virtue of which kinds are defined we can never be sure whether our differentiations match those kinds. The usual solution, as is well known in its contemporary linguistic form, is simply to claim that our

modes of differentiation (our use of terms) literally defines the range of their application. As I hope to show, this entails sacrificing a workable theory of truth. Nevertheless, I am sensitive to the requirements of verification, and I intend this work to point to a way of accommodating both the demands of truth and that of epistemic anthropocentrism.

39. If the capacity for reference is to entail proper use of a name in most situations, then this requirement does not seem to me too stringent. Consider the case of referring to an island of which "that" (pointing) is a beach. Without some knowledge of the boundaries of the object, one would not know that it was an island to which one thereby referred. Even "the landmass of which that is a beach" presumes some ability to identify landmasses. Without these (prior) abilities, one could at most refer to the beach: except, of course, that even the beach has (fuzzy but real) boundaries; it would not be wise to reduce the statement to "the beach of which this is sand" unless one has the stomach for regress.

40. There is no contradiction here: one acquires (or, at least can acquire) the knowledge necessary for any practical task in the process of learning and mastering the practice in question.

41. Section 4.6 is dedicated to giving an account of our access to sortals which avoids scheme/content distinctions.

42. It is worth stressing that the fact that there are inevitable limitations of our epistemic access to the world, imposed by such matters as the spectral range of our eyesight and hearing, is not by itself sufficient to imply that the world has no content-shaping influence on our thoughts, which is the necessary prerequisite for radical and undetectable incongruence between our thoughts and the world. Our conceptions of the world will always "fall short" of it in the non-radical sense urged on us by a humble appreciation of our human limitations; this is not at all what I am at pains to dispute.

43. Whether this mode be sensation-based descriptions or Frege's more complicated and difficult to understand senses, whose contents are apparently not restricted to perceptually available properties.

3
Pragmatism, Realism, and the Fourth Dogma of Empiricism

3.1: Progress, Truth, and Epistemic Openness: Peirce and James

"The truth," James famously wrote, "is what works." It is a peculiar claim. For like Dr. Johnson's refutation of idealism it appears to be a criticism which runs the opponent through while missing everything vital, indeed without even wounding. Of course the truth works—it is, after all, the truth. Surely one would expect accurate information to provide grounds for successful action. But there is something deeper afoot here. Contained in this laconic gem is an excoriation of the 'copy' theory of truth (perhaps the most familiar legacy of the pragmatic movement in philosophy): The true is not that which pictures the world accurately; rather it is that which usefully orients the agent, disclosing the world not as a vision, but as a possibility for action. Thus knowing the world is a matter not of the accuracy of the present presence of the world, but of the success of future endeavor. The truth *works*—that is both *succeeds* and *labors*—for the truth is not some characteristic of the relation which obtains between the mind of the knower and the world, but is rather used in (and is in part the activity of) relating the knower to the world.

Pragmatism, then, is a necessarily forward-looking doctrine, according to which the proof of knowledge claims is not a matter of current accuracy but future activity. As J. P. Diggins rather too critically notes, "Oriented toward the future, unburdened by the classical problems of knowledge and truth, rejecting the perished past as incapable of being represented (made present) in thought . . . pragmatism advises us to try whatever promises to work and proves to be useful as the mind adjusts to the exigencies of events."[1] There is a somewhat Darwinian mood to this philosophy, where the ultimate standard of success is the capacity to cope with and adjust to the future's continuous arrival in the present. If humans are to be considered naturalistically, it seems the only important standard for our knowledge to meet (indeed, the only standard for which we could have evidence for its success in meeting) is its usefulness to our survival. Failure occurs, to use Henry Adams' lamentation, when "experience ceases to educate."[2]

But in being educated, do we thereby come to understand better the actual world, or are we rather learning, quite literally, to make a better world? Although the distinction is of dubious value when applied to social thought, it is still a powerful tool for thinking about the status of our claims to know the natural world. Richard Rorty, leaning heavily on Dewey and bolstered by Kuhn and Davidson, insists that the latter is the only genuinely pragmatic attitude. For Rorty, "pragmatism councils adjustment and adaptation on the assumption that ideals, or at least something of meaning and value, are expected to emerge from experience and hence need not be regarded as external to it."[3] Indeed, for Rorty pragmatism should be hailed chiefly for giving up on the idea that there *is* anything external to thought or discourse which we must strive to mimic. The choice is stark: solidarity or objectivity—and "pragmatism [is] a philosophy of solidarity."[4]

> Those who wish to ground solidarity in objectivity—call them "realists"—have to construe truth as a correspondence to reality. So they must construct a metaphysics which has room for a special relation between beliefs and objects which will differentiate true from false beliefs. They also must argue that there are procedures of justification of belief which are natural and not merely local. So they must construct an epistemology which has room for a kind of justification which is not merely social but natural, springing from human nature itself, and made possible by a link between that part of nature and the rest of nature. On their view, the various procedures which are thought of as providing rational justification by one or another culture may or may not really *be* rational. For to be truly rational, procedures of justification *must* lead to the truth, to correspondence with reality, to the intrinsic nature of things.
>
> By contrast, those who wish to reduce objectivity to solidarity—call them "pragmatists"—do not require either a metaphysics or an epistemology. They view truth as, in William James' phrase, what is good for *us* to believe. So they do not need an account of a relation between beliefs and objects called "correspondence," nor an account of human cognitive abilities which ensures our species is capable of entering into that relation. . . . From a pragmatist point of view, to say that what is rational for us now to believe may not be *true*, is simply to say that somebody may come up with a better idea. . . . For pragmatists, the desire for objectivity is not the desire to escape the limitations of one's community, but simply the desire for as much intersubjective agreement as possible. . . . Insofar as pragmatists make a distinction between knowledge and opinion, it is simply the distinction between topics on which agreement is relatively easy to get and topics on which agreement is relatively hard to get.[5]

For the pragmatist, then, "the consensus of a community rather than a relation to nonhuman reality is taken as central" to truth, and notions like 'better' for ideas and theories, and 'progress' for science and society can be given sense only in terms of criteria *internal* to the community in question: better for some particular purpose, progress toward some particular goal.[6] "To say that *we* think we're heading in the right direction is just to say, with Kuhn, that we can, by hindsight, tell the story of the past as a story of progress."[7] In particular, 'better' cannot justifiably mean that we are getting closer to the (unrepresented, unthought) world, for even if it makes sense to believe in such a world, it makes no sense to measure one's progress against it, for it is, *ex hypothesi,* the unthought, unrepresented, and therefore inaccessible world.[8] It has become, as I had chance to note briefly in the introduction, a platitude of contemporary philosophy that percepts without concepts are blind; insofar as knowledge is perceptual, there can be no knowledge of a "pure" unconceptualized world.

Nevertheless, somewhat to Rorty's embarrassment, there is room in the pragmatic movement for the conviction that it *is* the world standing outside of experience which educates us, through experience, to better understanding and better science. For Peirce especially, and to some lesser extent for James "the fact that diverse thinkers agree in a common result is not to be taken simply as a brute fact; on the contrary, the convergence of many observations, ideas, views in a common object stands in need of explanation."[9] These early pragmatists could accept an analysis of truth in terms of community agreement without thereby giving up on realism in science, and yet could maintain their realism without relying on a notion of truth as correspondence. For Peirce, "this great hope is embodied in the conception of truth and reality. The opinion which is fated to be ultimately agreed to by all who investigate is what we mean by the truth, and the object represented in this opinion is the real."[10] Likewise James: "the 'absolutely' true, meaning what no further experience will ever alter, is that ideal vanishing point towards which we imagine that all our temporary truths will some day converge."[11]

These notions of fated agreement and convergent opinion depend on twin posits: there is a world which exists independently of thought, and experience gives us access to this world. Without a world there would be nothing to guide convergence, and without access there would be no way for the convergence to be guided. Although we must be ever mindful of our own limitations, and be ready tomorrow to call false what we today call true,[12] still if we are vigilant and investigate with care and discipline, the world can act to limit and direct our inquiry. As James aptly puts it: "Experience, as we know, has ways of boiling over, and making us correct our present formulas."[13] Here truth can be simultaneously connected with (convergent) community opinion *and* understood in terms of its appropriateness to the (actual) world, for there are resources within early pragmatic-empiricism for the world to "show through" our conceptions of it, revealing the discrepancy between belief and reality. It is part of the purpose of this chapter to see if this middle ground can still be held.

Still, the thesis I am here defending can be further narrowed, for central to any realism is the notion that the content of, and changes in, theories and concepts can be attributed to the world; this idea was certainly central to the future-oriented, fallibilist realism of Peirce and James.[14] Likewise, denying that citations of the world[15] have explanatory value in understanding theory change is a staple of any idealism, just as it is for Rorty's pragmatism.[16] Which side we take in this disagreement depends entirely on the resources of our account of experience, for only if that account allows for the possibility that the world can show through whatever contextual limitations are imposed on an investigator by her language, culture, beliefs, or other aspects of her conceptual apparatus, can we make room for the idea of the world as guide. I will argue here that such an account of experience can be given if we take seriously the active body as itself a mode of epistemic access, possessing a nonperceptual openness to the world.

But to make this argument we must first understand the pragmatic-empiricist account of experience, and why it fails to ground the convergent realism to which it aspires.

The essence and root of pragmatism, Peirce once confessed, lies in Alexander Bain's definition of belief: that upon which a man is prepared to act. "From this definition, pragmatism is scarce more than a corollary."[17] This conception leads in short order to a proto-functionalist theory of belief content, thus:

> The essence of belief is the establishment of a habit; and different beliefs are distinguished by the different modes of action to which they give rise. If beliefs do not differ in this respect, if they appease the same doubt by producing the same rule of action, then no mere differences in the manner of consciousness of them can make them different beliefs, any more than playing a tune in different keys is playing different tunes.[18]

It is this reformulation of the nature of a belief that grounds the critique of truth briefly outlined above, which, it is important to note, was meant in part as a way of avoiding the trap of Cartesian skepticism without having to settle for coherence as the only meaning for truth[19]. If the content of belief was to be parsed in terms of some "copy" of the way the world looked, any justification of a particular belief (in the absence of a benign Creator and a guaranteed Method) would necessarily rely on a capacity to compare the content of the belief with some privileged set of perceptions or intuitions (which purported to represent the world as it "really" is).[20] In the normal course of events the test of truth is coherence, in this case coherence with the "privileged" representations; indeed, outside of coherence there is no "test" of truth at all, for the privileged representations are not (cannot be) in need of justification or testing. Since there is some set of representations providing uniquely accurate access to the world, such a system of justification (as does Descartes' own) allows coherence to be the test for truth, while retaining correspondence as the meaning of truth. But if it is determined that we have no capacity which allows us to determine some set of representations as unproblematic, coherence must exhaust the *meaning* of truth as well, for we lose our justification for supposing any particular set of representations, however coherent, to match the world.

Peirce seeks to avoid this conclusion by altering the content of the beliefs in need of justification, thus: "If one can define accurately all the conceivable experimental phenomena which the affirmation or denial of a concept could imply, one will have therein a complete definition of the concept, and there is absolutely nothing more in it."[21]

Or, as James would put it: "Beliefs, in short, are really rules for action; and the whole function of thinking is but one step in the production of habits of action. If there were any part of a thought that made no difference in the thought's practical consequences then that part would be no proper element in the thought's significance."[22]

The essence of a belief was to be understood in terms not of propositional content but of activity. By denying the importance to belief of "phenomenological" or "appearance"-based content and instead identifying beliefs with habits of action, a method of justification might be developed such that determinations of phenomenological accuracy were not necessary for confidence in one's beliefs. For Peirce:

> The idea or sign with which the cognitive process begins must have consequences. . . . A knowledge claim must always have a warrant to the effect that the original sign expresses something about the intended object that can be found in the object if we follow where the sign leads. The process taking us beyond the initial idea is the process of critical testing, only from the outcome of this process can we learn whether the intention was fulfilled or disappointed. The cognitive move is always from intention to outcome, truth means intention fulfilled and error is intention disappointed.[23]

Beliefs are never merely representations of how things are taken to be, but are always also expectations regarding how situations will develop under certain conditions. "Whatever assertion you may make . . . [it] will either [be understood] as meaning that if a given prescription for an experiment ever can be and ever is carried out in act, an experience of a given description will result, or else [it will make] no sense at all."[24]

Thus justification need not rely on the comparison of normal experience with some set of experiences definitive of the state of the world, for a belief did not purport to define that state directly. Rather, one's knowledge of the world was anticipatory, a sense for "what sensations we are to expect from [the world], and what reactions we must prepare."[25] If, when, and for as long as these expectations were fulfilled, when the world reacted as supposed, the belief could be considered justified.

Although it is not part of this chapter to inquire as to the success of the pragmatists in avoiding a particularly Cartesian skepticism, it is nevertheless necessary to ask about their warrant for supposing that truth (justified belief) could still be understood as a kind of appropriateness to the world. What was the ground for supposing that the coherence of intention (expectation) and experienced result could indicate a "fit" between their knowledge and the world?[26]

The question can be subdivided: under what conditions (or for what reason) would it happen that increasing the coherence of our knowledge (and thereby, for the pragmatists, the reliability of our expectations) would imply increasing appropriateness to the world itself? And what form could this "match" between thought and world take, given the pragmatic critique of representation?

The answer to the second of these questions comes in noticing the pragmatist's assimilation of the "modern" reconceptualization of nature. The objects of the world, and the world itself, were coming to be understood not in terms of timeless essences, but in terms of actions, reactions and interactions. Thus "science need not grasp the inherent qualities of things but instead [need only] study its relations and connections with other things."[27] If the content of one's concept of a particular object consisted in a series of expectations for how, and under which conditions, it would react, and if the objects of the world could be understood themselves on the same model, this opened the possibility of a different kind of "fit" between concept and object.[28]

But the first question is more difficult for the fallibilist Peirce and James, for they adopt a coherentist, historically oriented method of inquiry and an anti-representationalist attitude (usually associated with idealism), but nevertheless aim at a true and complete description of the world. Yet once we accept the idea that there is to be no immediate, infallible access to the nature of things there are two sorts of limitations to human inquiry which we must acknowledge: the individual thinker is *perceptually* fallible, and any community is *historically* fallible. For Peirce, science was the solution to the limitations of the individual, for the confluence of perceptions and opinions of the many could be taken as reliable evidence that there was *some* publicly available object under investigation which could be cited as a reason for the observed agreement, and this ruled out at least private hallucination or simple mistakes. As John Smith notes:

> Peirce believed that it was legitimate to take the idealist line and connect reality essentially with thought and opinion as long as the thought involved was independent of the finite and fallible thinker, and as long as it was not conceived in a "mentalistic" way. The interpretation of conceptual meaning in terms of habit and behavior rather than intuited, immediate ideas is intended to satisfy the latter demand, and the community of inquirers following an interpersonal method is supposed to satisfy the former demand.[29]

Further, the rigorous method of science "requires individual inquirers to constitute themselves as members of the community of science through their willingness to sacrifice their privacy and bind themselves to the rules of an interpersonal method."[30] The members of any community of knowers can provide a test of the perceptions of the single individual; science was special in that it also purported to make the individual herself more reliable.

It is not, however, clear how Peirce hopes to mitigate the historical fallibility of the scientifically constituted community. As noted above, Peirce imagines that a well-constituted community is "fated" to converge on the truth, but we have from him no clear account of the mechanism of fate.[31] This lacuna in his work should, I think, be attributed to Peirce's own attraction to a kind of idealism, wherein it would be mistaken to define reality as existing entirely apart from thought. He writes that the real is "that whose characters are independent of what anybody may think them to be," but it is not clear that reality is independent of what *everybody* may think it to be: "reality is independent, not necessarily of thought in general, but only of what you or I or any finite number of men may think about it."[32] Thus we have Smith's valiant effort:

> It is not consistent with the realistic passages in Peirce's writings to say that for him knowledge constitutes reality . . . on the other hand, reality is defined through opinion and thought. . . . The fact is, and this is the crux of Peirce's entire philosophy, he believed it defensible to connect reality essentially with thought as long as the thought in question is of a certain ultimate character that does not depend upon finite thinkers who taken singly are fallible and without final authority. The thought or opinion that defines reality must therefore belong to a community of knowers, and this community must be structured and disciplined in accordance with superindividual principles.[33]

Except for the passing reference to the "ultimate" character of thought, Smith has unified Peirce's thought only at the expense of jettisoning his convergent realism in favor of a more thoroughgoing intersubjective idealism. Indeed, if we take language to be a prime instance of the "superindividual principles" in terms of and by which the community is to be constituted, this passage makes an apt description of the linguistic-idealist tendency in Wittgenstein. But as Smith himself notes, this reading, which implies that community opinion "defines" reality, is simply not consistent with Peirce's professed realism.

I think we can give a better accounting of Peirce's intentions, if not his actual expressions, in the following way. In accordance with Peirce's coherentism, reality must be *known* in terms of the concepts, language and standards of one's community; knowledge of the real can be essentially connected with thought, without this having to imply that the real is defined by thought. To accept our historical fallibility is to recognize the fact that no current set of beliefs can be considered definitive, precisely *because* the real is independent of thought.[34] But if this is to be accepted without implying skepticism, and overcome without implying idealism, the burden of justification must be placed squarely on the future. It is for this reason that the pragmatists must embrace the possibility of progress, and must be able to define progress precisely in terms of improving accounts of the world itself.[35] For although perhaps there is not now a perfect "fit" between belief and reality, if it can be supposed that continuous inquiry can, in revealing our mistakes, constantly improve our conception of the world, then we would be justified in equating perfect coherence, that state which "no further experience will ever alter," with (non-representational) correspondence. And as Peirce himself (occasionally) recognizes, it is only by standing outside of and unconstituted by community opinion[36] that the world can act as the object of investigation which could *lead* community opinion.

But this reconstruction places a heavy burden on Peircean pragmatism, for once we have accepted the bifurcation of knowledge and reality we must face the question, as with all dualisms, of the relation of the two. We need to give an account of how the world can lead opinion, and how opinion can be led. The possibility of a convergence to some "fit" between the world and community opinion must be understood, I think, in terms of an account of epistemic access, and any account of epistemic access must focus on the individual knower. For it is only in virtue of the individual acting as a kind of mediator between the accepted opinion and shared expectations of the community and the world itself, that the world has "access" to opinion to influence it, and the community has access to the world to know it.

This needn't imply that the individual is unmediated in her contact with the world: for Peirce, as we have noted, the knowing individual, as a condition of knowing, must internalize a socially instantiated system of concepts, habits, and methods. Thus it would be just as well to say that the community, in a kind of conceptual facsimile, mediates the relation between individual and world. This mediation is of course necessary as a way of overcoming the fallibility and privacy of the individual knower, and illuminates the conceptual impossibility of a radical epistemic individuality. But although the reliability of the individual depends upon his membership in a community, the reliability of that community's opinion depends equally upon the capacities of that individual. For if there *is* a world external to the community, it is the individual who occupies, copes with and contacts it.[37] Any justification of convergent *realism* will therefore ultimately rest on an account of (the structure of) individual experience which allows the world to act as the *telos* of inquiry.

James is the place to turn for this account. James analyzes the influence of society and history on the individual in terms of the storehouse of accepted beliefs and prejudices which the individual brings to her encounter with the world.

The individual has a stock of old opinions already, but he meets a new experience that puts them to a strain. Somebody contradicts them; or in a reflective moment he discovers that they contradict each other; or he hears of facts with which they are incompatible; or desires arise in him which they cease to satisfy. The result in an inward trouble . . . which he seeks to escape by modifying his previous mass of opinions. He saves as much of it as he can, for in the matter of belief we are all extreme conservatives. So he tries to change first this opinion, then that (for they resist change very variously), until at last some new idea comes up which he can graft upon the ancient stock with a minimum of disturbance of the latter.[38]

For James the movement of our opinion is something of a dialectical process: we are confronted with a contradiction and we search for a new idea, theory or framework which will best make coherent the new experience and the old. Our historical limitation is largely a function of the fact that our thinking has a great deal of inertia to it: having adopted a direction it is difficult to move us off course. But acting as a heavy keel is not the only way in which our past beliefs influence future direction: the content of the newly adopted idea (and perhaps therefore even our eventual interpretation of the "contradictory" experience) will itself be influenced by the content of past belief. That is, it is not just that the ship is hard to turn given the small force of a new idea, but given the influence of past experience the new idea is not likely to call for much of a change.

[When] the new idea is adopted as the true one . . . it preserves the older stock of truths with a minimum of modification, stretching them just enough to make them admit the novelty, but conceiving that in ways as familiar as the case leaves possible. An *outreé* explanation, violating all our preconceptions, would never pass for a true account of a novelty. We should scratch around industriously till we found something less eccentric. The most violent revolutions in an individual's beliefs leave most of the old order standing.[39]

With insight which nods in the direction of Kant and anticipates the perceptual psychology of Quine, Kuhn and Davidson, James argues that the novelty which gives rise to the irritation of doubt is itself understood in as familiar terms as possible. And yet for James it is still clearly possible for the novelty to cause a change in our terms and theories, to force us eventually to break out of old molds. As he aptly put it, "experience has a way of boiling over" when action does not lead to the expected result.

The notion that experience has the capacity to allow the world to "show through" our current conceptions of it is the result of James' curious combination of acceptance and critique of the traditional empiricist model of experience. James could be called, after the title of a collection of his essays, a "radical empiricist:"

According to his view the account of experience advocated by Hume and Mill is also found to be inadequate, but not because experience needs to be supplemented by a transcendent reason; on the contrary, James' charge is that classical empiricism was not empirical enough and so failed to note all that experience contains.[40]

As should not be surprising given the pragmatic notion of belief content already outlined, "experience" is not to be conceptualized as a set of neutral, unstructured, atomic sense data waiting to be given comprehensible form by theories, concepts and/or habits of association; rather, experience is itself rich enough to be the source of the expectations which ground belief. There is within current experience a sense of the movement of the now into the future, which is provided not by any organizing theory, but by the experience itself.

> Experience, in short, contains transitions and tendencies. . . . Nouns and adjectives are not the only names for the contents of experience; prepositions, copulas, and conjunctions are also required if we are to have a faithful analysis of all that experience delivers.[41]

This is at once an acceptance of the now popular understanding that experience is always holistic, that the idea of sense-data must be replaced with the notion of a temporally instantiated sensory field, and a denial that this coherent totality must be the result of some unifying conceptual hierarchy by means of which experience, as comprehended, is possible. Instead the terms and structure of this unity arise from experience itself: "Prepositions, copulas and conjunctions, *is, isn't, then, before, in, on, beside, between, next, like, unlike, as, but*, flower out of the stream of pure experience, the stream of concretes, or the sensational stream, as naturally as nouns and adjectives do."[42]

As Smith notes, this need not (as we shall see, cannot) imply that *all* relations flower from pure experience; indeed, even within the context of our present inquiry we can see the inadequacy of that view. For if all expectation were contained in experience itself, it is hard to see how we should ever be surprised; the fact that we are can only indicate (1) that we can misread experience (and thus that experience needs interpretation after all), and/or (2) that some of our expectation comes from a theoretical comprehension of the world which involves imposing on experience conceptions (and thus expectations) *external* to the sensational stream itself, and/or (3) that the transitions embodied in experience are not in fact accurate guides for the transitions which the world (or the world as experienced) will undergo.

Regardless of which combination of these three options we choose, it highlights the need for there to be some external relations uniting the stream of experience into a fully comprehensible whole.[43] For the third option means essentially that only the "atomic" content of experience is to be trusted, so that the reliable unity of the sensory field must be provided by the imposition of external associations (essentially the position of the traditional empiricist), while the second (closest to what I take James' position to be) asserts that the entirety of our epistemic relation to the world can be understood only by realizing both the full extent of the contents of experience itself, and also the necessity of that supplement to experience which thinking provides.[44] The first option is of course closest to our present perceptual psychology, inherited from Kant via Wittgenstein, whereby experience comes to have content only in virtue of its interpretation. In the end, I think that even James' empiricism cannot escape this first road, but we will deal more closely with this in the following sections.

For now it is important only to see that James' critique of empiricism still leaves (as he himself would have predicted) a great deal of agreement untouched. Two elements of this agreement are important to us here: the notion that all epistemic information is (ultimately) sensible, and the conviction that the theoretical structuring of experience still leaves us access to uninterpreted experience, which may therefore operate as a guide for the imposed external associations and conjectural theoretical structures of scientific and quasi-scientific investigation.[45]

We can easily see how the latter conviction could be central to a faith in convergent realism. As we have noted, if all expectation came from experience itself, then insofar as experience was reliable we would not be surprised into correcting any beliefs; and if we *were* confronted with an anomaly it is not at all clear that we could change anything to improve the challenged belief. For were our beliefs to depend for their content entirely on experience alone, any improvement in the reliability of those beliefs would depend on the improvement of our experience; this is to be hoped for but is not something which could be controlled, expected or even recognized. Likewise, if all expectation came from some set of organizing theories and concepts, and we thereby lacked the power to see how experience meant *itself* to be organized, that is, if the actual content of experience had no way of showing through our theoretical construct, then we should never be surprised by any discrepancy between the organized sensory field and the stream of pure experience. It is only when some important measure of our beliefs and expectations are rooted in theoretical understandings of the world (and thereby conceptual organizations of experience), *and* when experience itself has content capable of countermanding the expectations of thought, that discrepancy can arise, and improvement therefore becomes a possibility.

Thus it looks as though if there can be maintained a distinction between experience and our conceptualized version of experience then the notion of 'improvement' may have some use in explaining alterations of our theoretical models. But the possibility of realizing such improvement is not yet sufficient to justify belief in a convergent realism. For any realism must involve the idea that our conceptions are getting closer to the world, and what we have said so far is only that we can come to better cope with experience. To justify the further claim that coping with experience is coping with the world would require that we can safely treat the experience on which we base our conceptions, and against which we measure our improvement, as providing reliable access to the world. Ironically, it is the very features of pragmatism which made it possible to describe so easily a convergence which now raise problems for realism; the very fact that experience has content of its own raises the possibility that this content fails to be appropriate to the world.

Peirce thought to rely on the community as a way of assessing the reliability of individual experience, but in this case the assurance misses the mark, for what we want to verify is not individual experience, but precisely the agreed-upon community experience on which its theories are based. According to a traditional, and ultimately skeptical empiricism, belief in the accuracy of our current picture of the world was justified only insofar as the images which we continuously collect remain in agreement with that picture. But if experience itself is in question, the coherence of experience is no reliable guide. The initial attraction of pragmatism is that it holds out the possibility that activity could provide a test of experience more reliable than phenomenological coherence. For the pragmatist it looks as though belief could be verified in a much more satisfying sense, as our activity in the world confirmed what the belief had led us to expect. But this is not, in the end, a sustainable picture of traditional pragmatism. For both Peirce and James, all our encounters with and conceptions of the world are understood on the sensible model: "It is utterly impossible," writes Peirce, "that we should have an idea in our minds which relates to anything but conceived sensible effects of things."[46] Likewise James, "To attain perfect clearness in our thoughts of an object, then, we need only consider what effects of a conceivably practical kind the object may involve—what sensations we are to expect from it, and what reactions we must prepare."[47] In the end, activity itself is understood in sensible terms, and not just because it emerges here that action was important primarily because it increased one's sensory exposure to the world. The pragmatists, driven by their conception of belief as sensible expectation, conceive the principle of individuation (and thus identity) for actions in explicitly sensible terms. To give Murphy's excellent summary of the fundamentals of (Peircean) pragmatism:

1) Beliefs are identical if and only if they give rise to the same habit of action.
2) Beliefs give rise to the same habit of action if and only if they appease the same doubt by producing the same rule of action.
3) Beliefs produce the same rule of action if and only if they lead us to act the same in the same sensible situations.
4) Beliefs produce the same rule of action if and only if they lead us to the same sensible results.[48]

Thus, actions are individuated in terms of which sensations they arise in response to, and which sensations they ultimately lead to. Far from being a test of sensation in its own right, for Peirce and James activity is merely a step in achieving a test of coherence unlike the traditional empiricist test only in terms of its temporalization.

Insofar as this is correct, pragmatism cannot justify a convergent realism. As important as the deemphasis of depiction and resulting temporalization of belief are, alone they cannot justify the assertion that in improving our organizations of experience we thereby increasingly approximate reality. It is true that Peirce and James avoid both solipsism and what we might call simple idealism (whereby immediate experience is identified with reality) by insisting on fallibilism, the necessity of a community of knowers, and the continuing possibility of empirical testing; but without some test for experience's appropriateness to the world the continued adjustment of belief to experience amounts to a mere prejudice in favor of the future. Insofar as we place the burden of justification entirely on the promise of the future, our progressive improvement becomes a tautological illusion; for we will have without justification defined as progress whatever accommodations we make to future experience. In defining truth as the end of such a process of accommodation while insisting on describing the process as a convergence on reality, the pragmatist maintains for the present the "realist's" gap between thought and reality only in the temporal limit to identify them. As John Smith acknowledges, "reality in the end for Peirce is future experience, and this is not enough."[49] The resulting convergent intersubjective idealism seems the inescapable conclusion to the empiricist-pragmatist line of thinking.

Although perhaps this treatment has been overly destructive, I hold out the hope that it has been instructive as well. One key to the failure of the pragmatists to ground a realism was their acceptance of the empiricist tenet that sensation itself has content sufficient to ground and correct our conceptions of the world. The realist must hold out for an account of the organization of experience in which the world can play a substantial role in that organization, but to avoid the skepticism to which metaphysical realism seems prone it seems the world must play this role without adopting the guise of contentful experience. Perhaps giving up this dogma which, in distinguishing experience and thought, allows the former its own autonomous content, will allow us to see our way past the skeptical distance it opens. The next section will take stock of the promise offered to realism by the perceptual psychology of Donald Davidson, according to which experience need not be understood as the passive given to which our thought must conform; for Davidson our senses are tools of the mind, capable of being sharpened and improved by reason's proper functioning. Insofar as this is so, might it not be possible to explain our improving capacity to perceive the world in terms of a capacity to "see" how the world urges experience be organized?

3.2: The World Well Lost?: Davidson and Rorty

I stressed in the last section the failure of pragmatism to ground a convergent realism. The argument was based on skeptical considerations: no case could be made that the experiences to which conceptual understanding converged revealed the world. In this section I would like to follow out a parallel critique, both as a way of showing more thoroughly why the pragmatist-empiricist model must fail to ground realism, and also to point in the direction of some possible solutions to fill the void which this failure leaves. The general outlines of this critique, coming out of the work of Quine, Sellars and Kuhn, are too familiar to need much elaboration. According to this recent tradition of thinking, the skeptical critique of empiricism can be understood (and eventually bypassed) by looking at it in the following way: Central to any account of perception is the idea that sense organs are mechanisms in the physical world which transmit information based upon the impulses which their causal contact with the world engenders; the passive distance which contentful experience must therefore have from its causal origin opens the door for the skeptical attitude, driving a wedge between experience and the world. For the information gained from any such causal mechanism must necessarily be *interpreted*, and as Quine points out, there is no reason to suppose that there are any external limits on our possible interpretive arrangements of surface irritations.[50] Whatever constraints we do experience in our own understanding of reality can be attributed to the ingrained conceptual preferences and prejudices which limit our interpretive imagination.

It is, of course, left open for us to insist, with James,[51] that it is experience *itself* that has the content which limits our interpretive arrangements, but for this to aid us in constructing a realist epistemology we would have to answer two questions which are yet awaiting satisfactory solution: How could sensation, qua physical *cause*, dictate its own interpretation? Why should we suppose such self-interpreting causes to reveal the world accurately?[52] The answer to the first question, at least according to the orthodox tradition of analytic philosophy this century, is that no such cause could exist. All data are subject to interpretation as a condition of being significant; nothing signifies except within a context providing the terms and protocols of that signification. In the case of "pure experience" (simple observation) what seems like self-signification can at best indicate an implicit interpretive context. The critique gives rise to the following dilemma, which forms the guiding trope of John McDowell's *Mind and World*. The significance accorded to experience as given is meant to provide a limit on possible conceptual overlays, but if experience is to provide a *reason* why it should be understood one way rather than another, then it seems as if the experience must already be possessed of conceptual content—it must have significance within the conceptual sphere. Thus we are forced to contend either that the experience gains its significance in *being* conceptualized (which is simply to accept the critique, and deny the premise of the empiricist-pragmatist model of knowledge) or that non-conceptual content can somehow limit the workings of our interpretive understanding. But any account of experience as the reception of non-conceptual content sufficient to limit conceptualization would have to rest on a non-conceptual capacity for comprehension of that experience which allowed sufficient interpretive access to that stream of experience to glean which specific limits should be placed on its translation into a conceptual space. Yet this capacity initially to interpret *non-conceptual* experience and properly structure it *conceptually* would have to operate without being subject to the same rules of interpretation which caused the original dilemma. It certainly looks as though any attempts to provide an account of experience along *these* lines would collapse into incoherence, which leaves us to suppose that experience limits conceptual overlays not in virtue of some *content* or conceptual significance, but precisely in virtue of its *causal* powers. But no cause can provide a justification for the limitations it imposes. As McDowell writes:

> What we wanted was a reassurance that when we use our concepts in judgement, our freedom—our spontaneity in the exercise of our understanding—is constrained from outside thought, and in such a way that some of our judgements can count as justified. But when we make out that the space of reasons is more extensive than the conceptual sphere, so that it can incorporate extra-conceptual impingements from the world, what we get is a picture in which constraint from outside figures at the outer boundary of the expended space of reasons, in what is depicted as brute force from the exterior. . . . What happens there is the result of an alien force, the causal impact of the world, operating outside the control of our spontaneity. But it is one thing to be exempt from blame, on the ground that the position we find ourselves in can be traced ultimately to brute force; it is quite another thing to have a justification. In effect, the idea of the Given offers excuses where we wanted justifications.[53]

And yet to reject the idea of the Given, to hold the apparently reasonable idea that the content of our *experience* is always conceptual, seems to offer no way to limit the possible conceptual arrangements of (what we take to be) reality; this threatens the very idea that we have an epistemic hold on the world at all. As Rorty notes:

> Since Kant, we find it almost impossible not to think of the mind as divided into active and passive faculties, the former using concepts to "interpret" what "the world" imposes on the latter. . . . But as soon as we have this picture in mind it occurs to us, as it did to Hegel, that those all important *a priori* concepts, those which determine what our experience or our morals will be, might have been different. . . . The possibility of different conceptual schemes highlights the fact that a Kantian unsynthesized intuition can exert no influence on how it is to be synthesized—or, at best, can exert an influence we shall have to describe in a way . . . relative to a chosen conceptual scheme. . . . Insofar as a Kantian intuition is effable it is just a perceptual judgement, and thus not merely "intuitive." Insofar as it is ineffable, it is incapable of having an explanatory function. This dilemma . . . casts doubt on the notion of a faculty of "receptivity." There seems no need to postulate an intermediary between the physical thrust of

the stimulus upon the organ and the full-fledged conscious judgement that the properly programmed organism forms in consequence.[54]

McDowell agrees with the diagnosis and prescription:

A genuine escape would require us to avoid the Myth of the Given without renouncing the claim that experience is a rational constraint on thinking. . . . we can do that if we can manage to conceive experience as already possessed of conceptual content.[55]

The line of thought which Rorty and McDowell exploit here comes directly from the work of Donald Davidson, who argues that any story which relies on a notion of experience as some set of neutral data waiting to be organized (which can therefore be organized in myriad different ways) is untenable. We have seen already that we cannot embrace the idea that experience is possessed of non-conceptual content which limits conceptualization, so we are urged to move forward to the idea that experience has content *only* in conceptualization—there is nothing which can be called experience which mediates between concepts and the world, experience simply *is* conceptual understanding. Davidson writes:

We have been trying to see it this way: a person has all his beliefs about the world . . . how can he tell if they are true, or apt to be true? Only, we have been assuming, by connecting his beliefs to the world, confronting certain of his beliefs with the deliverances of his senses. . . . No such confrontation makes sense, for of course we can't get outside of our own skins to find out what is causing the internal happenings of which we are aware. Introducing intermediate steps or entities into the causal chain, like sensations or observations, serves only to make the epistemological problem more obvious. For if the intermediaries are merely causes, they don't justify the beliefs they cause, while if they deliver information, they may be lying. The moral is obvious. Since we can't swear intermediaries to truthfulness, we should allow no intermediaries between our beliefs and their object in the world. Of course there are causal intermediaries. What we must guard against are epistemic intermediaries.[56]

If we endorse scheme/content divisions, it begins to look plausible that another person could have an interpretation of the world radically different from our own. But if we give up on the idea, we must also give up on the possibility of such radical difference, and not just because there is nothing which two believers could radically differ *about*. Davidson argues that limitations on the possibility of interpretation rules out global divergence in our understanding of the world. He imagines a case of the following sort: A friend exclaims, in response to a ketch sailing by the dock where you both are standing, "What a handsome yawl!" If we are going to be able to make sense of the statement, then we need to make some interpretive adjustments.

One natural possibility is that your friend has mistaken a ketch for a yawl, and has formed a false belief. But if his vision is good and his line of sight is favorable it is even more plausible that he does not use the word "yawl" quite as you do, and has made no mistake at all about the position of the jigger on the passing yacht.[57]

If we are to get a foothold on understanding what any interlocutor is saying, and what he is saying it about, then the interpretation needs to proceed according to the principle of charity; our interpretation of his words depends upon our attribution to him of a set of largely true beliefs. And this means attributing to the other much of the set of beliefs which you yourself hold.

Such examples emphasize the interpretation of anomalous details against a background of common beliefs and a going method of translation. But the principles involved must be the same in less trivial cases. What matters is this: if all we know is what sentences a speaker holds true, and we cannot assume that his language is our own, then we cannot even take a first step towards interpretation without knowing or assuming a great deal about a speaker's beliefs. Since knowledge of beliefs comes only with the ability to interpret words, the only possibility at the start is to assume general agreement on beliefs. We get a first approximation to a finished theory by assigning to sentences of a speaker conditions of truth that actually obtain (in our own opinion) just when the speaker holds these sentences true. . . . We must conclude, I think, that the attempt to give a solid meaning to the idea of conceptual relativism, and hence the idea of a conceptual scheme, fares no better when based on a partial failure of translation than when based on total failure. Given the underlying methodology of interpretation, we could not be in a position to judge that others had concepts or beliefs radically different from our own.[58]

But it is not clear that this artful dodge of the problem of relativism actually furthers the case for realism, a condition for which (as I have argued) is the capacity to attribute the content of one's beliefs to the world, whether indirectly by attributing the latest conceptual refinement to the world, or directly by finding some ground in the world for the very concepts in terms of which experience is cast. I noted late in the last section that collapsing perception and conception might offer the realist the grounding for a (convergent) realism which the separation of the two failed to provide. Can we find the resources from within this perspective to make sense of the claim that the world can improve our perception of it?

The answer coming from Davidson must be no, although Davidson himself does not give up on objectivity. For Davidson the fact that "belief is in its nature veridical"[59] is itself sufficient to guarantee that beliefs are about the world; true beliefs could hardly fail to be about the world. But Davidson thinks that this aboutness is the natural result of the truth and coherence of our belief system, itself the result of the conditions determining the "existence and contents of belief."[60] Davidson writes:

> [T]ruth is correspondence with the way things are. . . . So if a coherence theory of truth is acceptable, it must be consistent with a correspondence theory. Second, a theory of knowledge that allows that we can know the truth must be a non-relativised, non-internal form of realism.[61]

Davidson thinks he has provided for such a theory of knowledge, which allows us to know the truth, and since truth is "correspondence with the way things are," it follows that what we know in knowing the truth is precisely the way things are. Of course, Davidson is perfectly aware of the traditional problems facing any coherence-based theory of knowledge and truth in claiming that the known is an objective world.

> If meanings are given by objective truth conditions there is a question how we can know that the conditions are satisfied, for this would appear to require a confrontation between what we believe and reality; and the idea of such a confrontation is absurd. . . .There is no way to get outside our language to find some test other than coherence.[62]

But for Davidson, this dilemma rests on the problematic notion that meaning is given by "objective truth conditions" by definition inaccessible to our concept-laden beliefs, and Davidson has shown that this theory of the basis of meaning is untenable. Without giving up on objectivity, he shows meaning is grounded in the fully epistemically accessible conditions of the interpretive situation.

> What concerns me here is that Quine and Dummett agree upon a basic principle, which is that whatever there is to meaning must be traced back somehow to experience, the given, or patterns of sensory stimulation. . . . once we take this step, we open the door to skepticism, for we must then allow that a very great many—perhaps most—of the sentences we hold true may in fact be false. It is ironical. Trying to make meaning accessible has made truth inaccessible.[63]

The problem is that "objective" was taken to imply "independent of interpretation," and this is just what opens the skeptical gap. For Davidson meaning is based precisely on interpreted experience, that is, concept-laden experience, because there simply is no other sort. Davidson aims to preserve both objective truth conditions and accessibility by locating experience within the space of concepts, and meaning in the intersubjective confrontation of the interpretive situation, thereby altering the sense of "objective" from "grounded in uninterpreted facts" to "indefeasible."

If coherence is a test of truth, then coherence is a test for judging that objective truth conditions are satisfied, and we no longer need to explain meaning on the basis of a possible confrontation. . . . Given a correct epistemology, we can be realists in all departments. We can accept objective truth conditions as the key to meaning, a realist view of truth, and we can insist that knowledge is of an objective world independent of our thought and language.[64]

But it is not at all clear that this side-step of the problem of relativism actually puts us in a position to attribute our knowledge to such an objective world, *"independent* of our thought and language." Richard Rorty, speaking at least partly in defense of Davidson, ridicules this desire for an objectivity grounded in independence:

[Davidson's] way of dealing with the claim "it is the world that determines what is true" may easily seem a fraud. . . . [The] view seems to perform the conjuring trick of substituting the notion of "the unquestioned vast majority of our beliefs" for the notion of "the world.". . . [But the realist] notion of the world is not a notion of unquestioned beliefs, or unquestionable beliefs, or ideally coherent beliefs, but rather a hard, unyielding, rigid, *etre-en-soi* which stands aloof, sublimely indifferent to the attentions we lavish upon it. . . . What [the realist] wants is the notion of a world so "independent of our knowledge" that it might, for all we know, prove to have none of the things we have always thought we were talking about.[65]

Of course this last bit of hyperbole is unfair:[66] what the realist wants is a world which *does* contain the things we are talking about, but which does not contain them *because* we are talking about them. And it is not at all clear that Davidson gives us such a world.

In the Davidson-Stroud position, "the world" will just be the stars, the people, the tables, and the grass . . . the fact that the vast majority of our beliefs must be true will, on this view, guarantee the existence of the vast majority of the things we now think we are talking about. So in one sense of "world"—the sense in which . . . we now know perfectly well what the world is like and could not possibly be wrong about it—there is no argument about the point that it is the world which determined truth.[67]

The problem here is obvious: given the relation Davidson imagines between beliefs and the world, one can very well say that the world determines truth, but the very same relation allows the statement that the truth of our beliefs determines the world. The realist needs this determination to operate in one direction only, but Rorty utilizes Davidson's position to ridicule this desire, which requires precisely the notion of objectivity as independence (rather than as solidarity) which Davidson argues we don't need and, insofar as it leads to skepticism, don't want. Rorty therefore embraces the ambiguity of the situation to argue convincingly that in Davidson's epistemology coherence (true belief determines world) and correspondence (world determines true belief) are non-competing trivialities; both are equally apt, and equally misleading characterizations of our epistemic position.

Davidson is not unaware of, and not entirely comfortable with, the indeterminacy to which his position apparently leads. We have seen already that he is happy with the idea that coherence and correspondence are mutually implicating, but he goes on to say:

> As Rorty has put it, "nothing counts as justification unless by reference to what we already accept, and there is no way to get outside our beliefs and our language so as to find some test other than coherence." About this I am, as you see, in agreement with Rorty. Where we differ, if we do, is on whether there remains a question how, given that we cannot "get outside our beliefs and our language so as to find some test other than coherence" we nevertheless can have knowledge of, and talk about, an objective public world which is not of our own making. I think this question does remain, while I suspect Rorty doesn't think so.[68]

The question does remain, and Davidson's epistemology takes us no farther than the indeterminate stalemate between coherence and correspondence which one can be forced to accept just because our epistemic position seems to offer no metaphysical alternative to skepticism.

But one is tempted to make a Wittgensteinian linguistic move: the idea that beliefs are guaranteed, that "*Noûs* cannot err,"[69] casts doubt on whether the word "right" or "true" has any use here. The theory of truth which Davidson/Rorty present cannot, it seems, live up to the standards of its own vocabulary. Of course Rorty has argued for some time that one result of his "pragmatic" investigations is the discovery that "true" is just the trivial term of praise which his theory reduces it to. But surely that revisionism whereby we take metaphysical-epistemic theories to supervene on everyday use is at least out of step with the spirit of the ordinary language philosophy which Rorty purports to inherit.

But in any case the charge is not entirely fair; certainly beliefs can be found false in the course of one's experience. The question for the realist evaluating Davidson's theory should be the following: in those cases where we revise our beliefs to cohere with current experience, can that adjustment be attributed to the world? It does not seem that Davidson's offers the resources whereby such an attribution is possible. Since no "confrontation" between belief and reality is conceivable for Davidson, any change in one belief can only be attributed directly to the adoption of another belief, for "nothing can count as a reason for holding a belief except another belief."[70] Since no experience-belief distinction can be maintained, there is no further attribution which can be made, except the trivial attribution to the world which Rorty allows. Any more robust attribution, one which cannot simply be reversed to the equally appropriate "my (new) world can be attributed to my belief" is blocked by Davidson's own conception of our contact with the world.

For Davidson, the world's influence is felt only causally, and because he has banished all content from such causes, no experience can be justified by reference to its cause. It is a necessary consequence of Davidson's theory that no cause can dictate the content of an experience, and this introduces an unbridgeable gulf between the content of experience—to which we attribute our change in belief —and its causal origin in the world. But if no experience can be substantively attributed to a cause, it is hard to see how it could be attributed to the world, if our only contact with the world is causal.

It is not my intention to deny this part of the argument: no cause can act as a justification for belief, and because all meaning requires a context of interpretation, the belief which any perceptual event causes cannot in any case be attributed simply to the nature of that cause. But I follow McDowell in thinking this threatens the very idea that beliefs are about the world.

> It does seem that the argument credited to Davidson by Rorty allows us to ring in changes on the actual environment (as seen by the interpreter) while things stay exactly the same from the standpoint of a believer, with the changes making no difference to the claim that interpretation reveals the believer as in touch with his world. This strikes me as undermining the claim that the argument traffics in any genuine idea of being in touch with something in particular. The objects that the interpreter sees the subject's beliefs as being about become, as it were, merely noumenal so far as the subject is concerned.[71]

Or, to paraphrase Wittgenstein, the wheel which can be turned though nothing moves with it is not part of the mechanism. Because the world conceived as only *causally* related to the believer via the sense organs can exercise no limitation on the content of the beliefs which the perceiver comes to hold, it would seem to follow that the world is not part of the epistemic mechanism.[72] This fact threatens the notion that we are in touch with a world—for surely it is a condition of having an epistemic link with some particular that the object in question exercise some limiting or guiding influence on the beliefs one is likely to hold in virtue of that link.[73] In this context, Davidson's assurance that we are automatically correct most of the time seems a bit hollow.

> Davidson's picture depicts our empirical thinking as engaged in with no rational constraint, but only causal influence, from outside. The result is to raise a worry as to whether the picture can accommodate the sort of bearing on reality that empirical content amounts to, and this is just the kind of worry that can make an appeal to the Given seem necessary.[74]

Without disputing Davidson's arguments that a reliance on some experiential Given gets us nowhere in philosophy, and certainly without making the claim that Davidson's epistemology implies that we could be "wrong" about the object of our beliefs, McDowell questions "how effectively the argument, operating as it does within Davidson's coherentist picture, can reassure us that the picture can really incorporate thought's bearing on reality."[75] In our blithe acceptance of the neo-Kantian dictum that percepts without concepts are blind, we forget its converse.

> Thoughts without intuitions are empty, and the point is not met by crediting intuitions with a causal bearing on thoughts; we can have empirical content in our picture only if we can acknowledge that intuitions have a rational bearing on thoughts. . . . I have suggested that we can do that if we manage to conceive experience as already possessed of conceptual content.[76]

3.3: McDowell and the Fourth Dogma of Empiricism
McDowell has set himself a difficult task. For we have seen already that *neither* giving intuitions their own content *nor* endowing them with causal powers can account for thought's bearing on reality, and reality's bearing on thought. And it is not at all clear what McDowell's insistence that "experiences already have conceptual content . . . in experience one takes in, for instance sees, *that things are thus and so*" adds to Davidson's version of the necessarily conceptual nature of perceptual belief.

According to McDowell, the judgement that things are thus and so, although it does not get to something non-conceptual, nevertheless

> takes us to something in which sensibility-receptivity is operative, so we need no longer be unnerved by the freedom implicit in the idea that our conceptual capacities belong to a faculty of spontaneity. We need not worry that our picture leaves out the external constraint that is required if exercises of our conceptual capacities are to be recognizable as bearing on the world at all.[77]

But McDowell has already ruled out anything which seems promising as a way of conceiving the mechanism of this constraint and the nature of our receptivity—and the mere fact that we seem receptive to some external constraint is hardly sufficient for the judgement that the world is doing the constraining. Thus it is not enough for McDowell to assure us:

> In experience one finds oneself saddled with content. One's conceptual capacities have already been brought into play, in the content's being available to one, before one has any choice in the matter. . . . In fact it is precisely because experience is passive, a case of receptivity in operation, that the conception of experience that I am recommending can satisfy the craving for a limit to freedom that underlies the Myth of the Given.[78]

Yet surely not all limits are good limits—the Evil Demon places as much a limit on our freedom in the skeptic's picture as the world purports to in the realist's. On what ground can McDowell make the claim that it is the *world* we are receptive to, which is the origin of the limits on the possible contents of experience which limits exist as a condition on the claim that we have an epistemic link with the world at all? How is it that "the idea of conceptually structured operations of receptivity puts us in a position to speak of experience as openness to the layout of reality?"[79] The problem facing McDowell is that he still has to elucidate the nature of the link between this conceptually structured capacity for accessing the world, and the world itself, and we have seen that a conceptually structured capacity of perception in *causal* contact with the world is inadequate.

> I find it helpful in this connection to reflect on a remark of Wittgenstein's: "when we say, and *mean*, that such-and-such is the case, we—and our meaning—do not stop anywhere short of the fact, but we mean: *this-is-so*." There is no ontological gap between the sort of thing one can mean, or generally the sort of thing one can think, and the sort of thing that can be the case . . . there is no distance from the world implicit in the nature of thought as such.[80]

McDowell's version of direct realism owes much to Aristotle (and no small amount to Wittgenstein's *Tractatus*). There is no *mediation* between thought and the world, nor need there be, for thought does not stop short of fact even in the ontological sense. This view hearkens to the Aristotelian idea that *logos*, the stuff of thought and language, is itself inherent in the world. Thus can the world call upon thought to grasp the world's natural structure, its *log*-ical form. "We should understand experiences in general as states or occurrences in which conceptual capacities are passively drawn into operation."[81] Having been deprived of causation as a means to understand our epistemic link to the world, the mechanism by which concepts are "drawn" into operation, McDowell seems to embrace a much more direct reception of nature, and to accommodate this possibility he is led to reconceptualize, as we might say, the nature of nature by relating it ontologically to thought. Not that McDowell is abandoning the lawlike generalizations of modernity's natural order, nor the idea that "natural science reveals a special kind of intelligibility."[82] Rather he is refusing to "equate the realm of law, so conceived, with nature, let alone the world."[83]

As far as I can tell, McDowell is endorsing—or at least indicating tentative support for—a capacity of Categorical Intuition, a natural sensitivity to the formal structure of the cosmos. It is hardly clear, for McDowell casts everything in the traditional vocabulary of perception—and perception, however conceptually structured, is generally understood to be *causal*; precisely the sort of link which has been shown to be epistemically inadequate. But although McDowell abandons talk of causation without giving up on perception, I think it unlikely that he (or we) would want to deny or theoretically sever all our causal connections with the world; our sense-organs are surely constituted by (and are epistemically important because of) their particular brand of causally operating receptivity. Nevertheless, any epistemology must work from within the recognition that these causal ties cannot in themselves do any epistemic work for us, and in particular they cannot influence the concepts brought to bear in perceptual experience.[84] In this context Categorical Intuition looks like just what the doctor ordered: a form of receptivity to the logical structure of the world sufficient to draw concepts into operation in our perceptual systems.[85]

There is no need to spend time in uncertain reconstructions of McDowell's intentions here; I do not know if this is a direction which McDowell would endorse, and I have no intention of following the line of thought to its conclusion. Allegiance to a form of Categorical Intuition has a distinguished history, and I have no immediate quarrel with that history.[86] What I wish to insist upon here is only that the solutions which seem open to McDowell (or anyone who, acquiescing to the arguments so far, still wishes for a theory according to which we *do* have some form of epistemically robust access to the *world*) all involve denying what I have called the fourth dogma of empiricism.[87] This dogma states in part that our only epistemic access to the world is perceptual—via the passive receptivity of the sense-organs.[88] For McDowell to find his way out of the problem he presents, he must endorse a form of openness to the world which goes beyond that exercised (causally) by our sense-organs, precisely because being in touch with the world implies that the content of experience will be restricted in a way that the causal operations of the sense-organs cannot explain.[89] And for his position to be a realism, it must be theoretically possible to attribute these limitations on experience to the influence of the world.

We have seen already that because experience is always already conceptual, whatever sort of openness to the world is meant to provide the limitations on the content of experience which realism demands, it will have to be capable of influencing the operations and structure of our *conceptual* system in just the ways which causal-perceptual receptivity cannot. Our claim to be in touch with the world depends upon our giving an account of experience which allows for the possibility that experience can be guided; in this context that means an account of the (direct) receptivity and openness to the world of that stock of empirical concepts which give content to experience.

Categorical Intuition is, of course, designed to fit just this bill. But it is clear that any such proposal would have its drawbacks, foremost among which would be the fact that it would involve supposing us to possess an epistemic capacity for which there is no direct evidence. Of course it is clear to the realist that we are in touch with the world and are right about at least the mid-size structure of the world—and the fact that perception cannot adequately account for the epistemic access by which we have such knowledge is no argument that we do *not* possess such epistemic access. But neither is the realist's conviction an argument for any particular account of our receptivity. Further, McDowell's account, even if it isn't the underpinning for a version of Categorical Intuition, involves a reconceptualization of the natural order (casting it in terms of thought) which may have undesirable metaphysical consequences. Although this is far from an argument against McDowell's work it does suggest that there is room for another way out of his dilemma. I have been slowly elaborating just such an alternative, one which utilizes a capacity we all readily acknowledge, the epistemic import of which is nevertheless ignored: the active body.[90]

In the following sections, then, I will return explicitly to an account of the epistemic resources of the active body; I will argue for the necessity of activity to any account of thought's bearing on (and direct reference to) material particulars, and give an account of the role of bodily activity in our acquisition and application of sortal concepts.

Notes

1.   John Patrick Diggins, *The Promise of Pragmatism* (Chicago: University of Chicago Press, 1994), 10, 2.
2.   Henry Adams, *The Education of Henry Adams* (New York: Modern Library, 1934), 294. Quoted in a similar context in Diggins, 17.
3.   Diggins, 20.
4.   Richard Rorty, "Solidarity or Objectivity?" in *Objectivity, Relativism and Truth* (Cambridge: Cambridge University Press, 1991), 33.
5.   Rorty, "Solidarity or Objectivity?," 22-3.
6.   Rorty, "Solidarity or Objectivity?," n1, 23.
7.   Rorty, "Solidarity or Objectivity?," 27.
8.   According to Rorty, for the pragmatists "it is useless to ask whether one vocabulary rather than another is closer to reality. For different vocabularies serve different purposes, and there is no such thing as a purpose which is closer to reality than another purpose." from the Introduction to John P. Murphy, *Pragmatism from Peirce to Davidson* (Boulder, CO: Westview, 1990), 3.
9.   John Smith, *Themes in American Philosophy* (New York: Harper & Row, 1970), 82. Peirce generally accepts as an explanation of consensus the notion that "external realities *cause* the common result and belief in one identical object." But see Smith for a discussion of Peirce's insistence that no simple causal theory of knowledge could be adequate.
10.   C. S. Peirce, *The Philosophical Writings of Peirce*, edited by Justus Buchler (New York: Dover, 1955), 38.
11.   William James, *The Will to Believe* (New York: Dover, 1955), 9. James is generally thought to be more "pragmatic-idealist" or phenomenalist than Peirce. Thus does Ellen Suckiel write "the most appropriate designation for his view as it applies to physical objects is that it is a proto-phenomenology: common sense objects as constituted by the

subject within lived experience are the ultimately real physical objects of our world." (*The Pragmatic Philosophy of William James* p.139) But William Gavin insists that James is a "radical realist" for whom language, as the "house of experience" points toward reality, a reality greater than and not entirely captureable by language. Nevertheless language, used properly, can give us access to reality, for, as Siegfried notes, "there is no gap between knowing and the known in pure experience." Gavin, *William James and the Reinstatement of the Vague* (Philadelphia: Temple University Press, 1992) and C. H. Siegfried, *Chaos and Context: A Study in William James* (Athens, OH: Ohio University Press, 1985), 111.

12. To paraphrase James (*The Will to Believe*, 9)

13. William James, *Pragmatism* (Indianapolis: Hackett, 1981), 100.

14. Simon Blackburn, in "Truth, Realism and the Regulation of Theory" singles out this disagreement as central to the division between realists and quasi- or anti-realists, although he questions its eventual usefulness in describing that division properly. I would like to suggest that the two positions have a tendency to collapse into one another at important junctures because realists and anti-realists are currently stuck in the same epistemic boat, which has the fourth dogma of empiricism as its rudder. (*Midwest Studies in Philosophy V: Studies in Epistemology.*)

15. Understood as indicating the world as it stands in contrast to how it is conceived. The denial that such citation is possible or useful takes many forms, of course, from the idea that reference is only possible in terms of, and via, concepts (so that no *reference* to an unconceptualized world is possible) to the (Wittgensteinian) claim that the limits of language describe the limits of the world (so that no sense can be made of the *distinction* between the world and a world conceived).

16. Davidson's famous argument, which we will deal with in some depth, is precisely a challenge to the idea (the third dogma of empiricism) "that we can distinguish changes in statements held true due to change in meaning [concepts] from those held true due to changes in beliefs [percepts]." (Murphy, *Pragmatism from Peirce to Davidson*, 99.) That is, it is a challenge to the idea that we can distinguish between attributing changes in statements held true to the world (because we had an anomalous experience/perception of the world) and attributing those changes to us (because we decided to use a word differently).

17. *Philosophical Writings of Peirce*, 270.

18. *Philosophical Writings of Peirce*, 229.

19. Here it seems that the *test* of truth and the *meaning* of truth might be different.

20. It is not at all clear that Descartes ever actually believed in a simple 'copy' theory of truth. See i.e., the Sixth Meditation (AT VII 79-86) and my "Certainty, Doubt and Truth," 28-30.

21. *Philosophical Writings of Peirce*, 253.

22. William James, *The Writings of William James*, edited by John J. McDermott (Chicago: University of Chicago Press, 1977), 348. But note that the theories of content proposed in these two quotes are significantly different. And this is not just a matter of disagreement between different authors: the Peirce quote on the previous page echoes the James quote above, and differs equally from the Peirce above. Indeed, it is arguable that pragmatism's own failure to take much note of this difference contributes to its inability to overcome the skeptical idealism with which it contends. As I will argue, it is precisely the pragmatist analysis of "habits of action" in terms of "experimental phenomena" which

prevents it from avoiding idealism.
23. Smith, *Themes*, 36-7.
24. *Philosophical Writings of Peirce*, 251.
25. *The Writings of William James*, 348.
26. It is of course easy to know when a belief is appropriate to experience. But the key to a realist position is to avoid a simple reduction of the world to experience. Else we end up where Quine puts us: "we have no reason to suppose that man's surface irritations even unto eternity admit of any one systematization that is scientifically better or simpler than all possible others. It seems likelier . . . that countless alternative theories would be tied for first place. Scientific method is the way to truth, but it affords even in principle no unique definition of truth." *Word and Object* (Cambridge, MA: MIT Press, 1960), 23.
  Certainly once we suppose that (1) experience itself ("surface irritations") has no content capable of resisting any organizing theoretical overlay and (2) our only epistemic input is via the experience described in (1), then we can hardly restrict the range of possible overlays, or possible truths, on epistemic or metaphysical grounds. James, as we will see, accepted (2) but denied (1). I hope to reinvigorate pragmatic realism by doing just the opposite.
27. Diggins, 13.
28. In chapter 4 I will argue in favor of the usefulness of this way of understanding the (potential) fit between scientific knowledge and the world: "One approaches the world by interacting with it, inserting oneself into the functional order, coming gradually to identify objects in terms of their place in the network of causal and equipmental contexts which we negotiate each day . . . interactions with the microscopic things of the world reveal for us their place in the extended causal networks with which scientists are familiar, and accepting their reality involves fitting them into that network which is ultimately continuous with the everyday." (From section 4.62).
29. Smith, *Themes*, 103.
30. Smith, *Themes*, 99.
31. Peirce toys with the idea of statistical destiny: "If we throw the dice often enough, sixes will be sure to turn up, although there is no necessitating reason why they should. By analogy this means that the real universe in which inquiry takes place is such that the ultimate opinion is sure to come about at some time, although there is no necessitating reason for this and there is no certainty that in fact the opinion has been reached at this or that particular time (see especially *The Collected Papers of Charles Sanders Peirce*, vol. 4 sec.457*n*.)" Smith, *Themes*, 10.
  This, of course, is entirely inadequate, and not at all in line with the idea that inquiry converges to a stable result—if we throw the dice after having gotten sixes (especially if we don't even know we have sixes!) there is no guarantee (even no likelihood) that they will turn up again. Peirce is more forceful about destiny, but less forthcoming about how it is possible, in the following: "The progress of investigation carries them by a force outside themselves to one and the same conclusion. . . . This activity of thought by which we are carried, not where we wish, to a fore-ordained goal, is like the operation of destiny." *The Collected Papers of Charles Sanders Peirce*, Vol. 5, sec. 407, quoted in Smith, *Themes*, 101.
  The question settles around the following problem: how can a force "outside of" a community of inquirers lead that inquiry?

32. C. S. Peirce, *Collected Papers of Charles Sanders Peirce*, Vols. 1-6, edited by Charles Hartshorne and Paul Weiss; Vols. 7&8, edited by Arthur Burks (Cambridge, MA: Harvard University Press), Vol.5, sec.405 & 407. Quoted in Smith, *Themes*, 83.
33. Smith, *Themes*, 91.
34. In a way, accepting historical fallibility (and not just the mere prospect of changing belief) implies a kind of realism. Hilary Putnam writes that "What does show that one understands the notion of truth realistically is one's acceptance of such statements as: (A) Venus might not have carbon dioxide in its atmosphere even though it follows from our theory that Venus has carbon dioxide in its atmosphere and (B) A statement can be false even though it follows from our theory (or from our theory plus the set of true observation sentences)." *Meaning and the Moral Sciences* (London: Routledge, 1978), 34.
35. Note that the account of progress need not imply the absence of "paradigm changes," although it does of course sometimes require that we describe such scientific revolutions as resulting in models and theories which are more appropriate to the world.

   Robert Almeder provides an account of Peircean fallibilism which relies specifically on the possibility of paradigm change: "the expression 'I know that p but I may be mistaken' need not *inconsistently* imply that p is true and possibly it is not. Rather it would simply mean that, given the semantic rules of our current conceptual framework, the assertion (or inscription) of p is authorized although it is *logically possible* that the conceptual framework in which p is presently authorized may be replaced by a future and more adequate conceptual framework in which the assertion (or inscription) of p may not be authorized." "Fallibilism and Ultimate Irreversible Opinion." *American Philosophical Quarterly* 9 (1975): 33-54.
36. That is, only if there is a difference between "being" and "being represented" can change in opinion be attributed to the (influence of) the world itself. As has been noted, Peirce was not at all consistent on this point, sometimes asserting that "the view of reality to which he is sympathetic is one that is "fatal to the idea of . . . a thing existing independent of all relation to the mind's conception of it." (Smith, *Themes*, 88.) and sometimes taking a more strictly realist view (see especially his critiques of Royce). Indeed, in his idealist moods, it is not even clear that it is the world to which opinion converges (but is rather a set of opinions of experiences), by which he appears to (sometimes) mean not the obvious that the final account of the world will of course be cast as an opinion, but that the opinion will in some sense constitute the real. An interesting hybrid of these two possibilities is given (in a line which is reminiscent of Putnam's internalism): "The opinion which is fated to be ultimately agreed to by all who investigate, is what we mean by truth, and the object represented in this opinion is the real." (*The Collected Papers of Charles Sanders Peirce*, vol.5 sec.407) I cannot claim, therefore, to have given an accurate, and certainly not a complete account of Peirce's thought. I think, however, that it is instructive in many ways, not just because of his attempt to come to terms with the strengths of both idealism and realism, but also because of the difficulties which this task presents. As I will argue in the next chapter, once we have accepted the idealist contention that thought (or conception) itself shapes experience (or perception), it is difficult to resist the lure of an objective idealism. Of course, as should be obvious by now, I hope to make the task easier (and thereby to place the pragmatists within a more hospitable epistemic-metaphysical frame than the empiricist one they have adopted) by arguing for the active body as a mode of epistemic access to the world.

37. Unless we imagine some version of an all-mind, constituted by the collective intelligence of the human community, to which the world is ultimately present, or accept a kind of idealism whereby this collective mind constitutes the world, the burden of our knowledge falls to the individual's contact with reality. Contemporary philosophy has tended to assimilate the community's apparent influence on knowing by positing the individual mind to be constituted by and in terms of a socially instantiated system of concepts and practices. I follow the contemporary line, but insist that this makes the reliability of community opinion depend upon the reliability of the experience of this socially-constituted knower.

38. James, *Pragmatism*, 31.

39. James, *Pragmatism*, 31.

40. Smith, *Themes*, 28.

41. Smith, *Themes*, 28.

42. William James, *Essays in Radical Empiricism* (New York: Longmans, Green, 1912), 95.

43. As Gavin points out, "James thought it totally unrealistic 'to hope that the mere fact of mental confrontation with a certain series of facts' would be 'sufficient to make any brain conceive their law.' The scientific observer was considered by James to be an active transformer of experience." Gavin, 67. James writes "The most persistent of outer relations which science believes in are never matters of experience at all, but have to be disengaged from under experience."

44. It is worth recounting, in this regard, Smith's critique of the possibility of finding all unifying relations in experience itself. "The fact is, and James himself was aware of the difficulty in the discussion of the a priori found in the final chapter of his *Principles of Psychology*, radical empiricism is unable to account for triadic relations on the basis of pure experience. These are relations on the basis of which two distinct terms, A and B, are related to each other in virtue of the fact that each is in turn related to the same third term C. Schematically this relation is expressed in conditional form, if A **R** C and B **R** C, then A **R** B. . . .

The first point is that when we have to compare two items we are unable to do so directly without the introduction of a mediating item or term. We do not compare two items merely by giving an exhaustive description of each. . . . no comparison takes place until there have been introduced relevant and specifiable respects or third terms to which each of the extreme terms can be intelligibly related.

If we attend to the nature of these third terms we can see that they do not present themselves as a part of the felt stream of thought after the fashion of conjunctive relations such as 'with', 'next', and 'between' which James finds in his pure experience. It is difficult to know what item in the stream of experience would be the 'original' for a relevant third term establishing a triadic relation required for comparison. . . . a significant third term required for comparison is an abstract term capable of generalization, and its *relevant* introduction into an act of comparison requires some ingenuity. . . . It might be thought that when we perceive a brown object and a blue one in close proximity and say that they differ, we are merely seeing a fact that is expressed in a comparative judgement. But this is not the case; the relevant respect of color has already been selected and applied. . . . Logically . . . the third term is always presupposed. [This is the case even on the model of aspect perception offered by Wittgenstein in *Philosophical Investigations* pt.II. -M. O'D-A.]

Chapter 3

The upshot of the foregoing discussion may be stated as follows . . . without the so-called house-born concepts and logical operations involving abstract and generalizable terms, there could be no systematic interpretation or explanation of experience. The stream of experience is neither self-organizing nor self interpreting. . . . Try as we will, we do not escape the problem of the a priori, the problem, in Kant's language, of the concepts that begin with experience, i.e., have empirical meaning, but do not arise out of experience in the sense that we can find no obvious 'originals' for them in the stream of direct experience." Smith, *Themes*, 37-40.

45. Nowhere do we see any indication in James of a non- or extra-sensory reception of the world. He does distinguish between 'perception' and 'sensation', but not in terms of their constituting different modes of access to the world, but rather in terms of their relative degree of *conceptualization*. Writes Gavin: "The closer an object cognized comes to being a simple quality like hot, cold, or red, the closer the state of mind comes to being a pure sensation. Perception, in contrast, includes sensation: 'sensation . . . differs from perception only in the extreme simplicity of its object or content. Its function is that of a mere acquaintance with a fact. Perception's function, on the other hand, is knowledge about a fact; and this knowledge admits of numberless degrees of complication.'" Gavin, 86.

Clearly sensation is the mode of epistemic access which James envisions. But although, as we have seen, pure experience (as far as I can see, 'experience' is a broader term than 'sensation', and includes conception, belief and perception, while 'pure experience' is synonymous with 'pure sensation') contains a great deal more than does the classical empiricist notion of (pure) experience, still our actual experience, when it is truly conscious, is never pure. Writes James: "For whom is 'pure experience' available? Only new-born babies." Thus, as Gavin notes, "pure sensation is an abstraction. All sensations are pre-perceived." (Gavin, 66.) Our actual experience is always a mixture of internal and external relations and conjunctions, with the internal relations ideally operating to limit the external ones.

46. *Philosophical Writings of Peirce*, 31.
47. *The Writings of William James*, 348.
48. Murphy, 25-6.
49. Smith, *Themes*, 107.
50. It is worth noting that this implies skepticism only on the assumption that knowledge is (quasi-)representational, and the definition of truth is correspondence. The account certainly quashes any hope that the interpreted data could correspond to an uninterpreted world, or that one could ever know that one's knowledge was accurate. To give up this sort of realism, then, for a coherence based intersubjective idealism is, it is supposed, to avoid skepticism.
51. The citation of James in this regard is meant only to emphasize his own insistence on the contentful nature of pure experience. I have argued that sensation was, for James, our sole mode of epistemic access to the world, but it is not clear that he thought of sensation as "causal," i.e., as a product of physical-causal mechanisms. No doubt James held that sensation was the product of sense-organs, which, being physical mechanisms must have transmitted their information causally; indeed, if he had some other mechanism in mind it surely would have been incumbent on him to make this clear (since in this case the obvious assumption would mislead).

Interesting in this regard is a remark of his on causation: "Hume's account of causation is a good illustration of the way in which empiricism may use the principle of totality; we call something a cause, but we at the same time deny its effect to be in any way contained in or substantially identical with it. We thus cannot tell what its causality amounts to until its effect has actually supervened." (James, *The Will to Believe*, 147) It may well be that, were we to follow out this line of reasoning, we would discover that there needn't be for James the usual problem of a "physical" cause leading to a "mental" effect—were a sensation or belief to result from the action of the sense organs, then that would be the determined nature of the cause.

52. Or, to put it differently, why should we suppose that it is the world which dictates the content of our interpretations? One research program which takes on these questions is evolutionary epistemology. The guiding intuition, too simply stated, is this: Our survival surely depends on our possession of a capacity to interpret appropriately the impulses of our sense organs. Those members of the species for whom certain causes have their proper significance will survive and pass on the genetics responsible for the capacity in virtue of which the match between sensory impulse and significance was achieved. It is interesting to note the similarity of the escape from skepticism which such accounts provide with Descartes' own suggestions. For both we are providing a story about our origins which is meant to equip us with the confidence that we have a capacity of knowledge which, when attended to properly, will not be deceived.

But see note 73 for some indications of the limits of the usefulness of evolutionary epistemology in this regard.

53. John McDowell, *Mind and World* (Cambridge, MA: Harvard University Press, 1994), I, 3.

54. Richard Rorty, "The World Well Lost," in *The Consequences of Pragmatism* (Minneapolis, MN: University of Minnesota Press, 1979), 3-4.

55. McDowell, *Mind and World*, I, 6.

56. Donald Davidson, "A Coherence Theory of Truth and Knowledge," in *Truth and Interpretation*, edited by Ernest LePore (London: Basil Blackwell, 1986), 312.

57. Donald Davidson, "On the Very Idea of a Conceptual Scheme," in *Inquiries into Truth and Interpretation* (Oxford: Oxford University Press, 1984), 196.

58. Davidson, "On the Very Idea of a Conceptual Scheme," 196-7.

59. Davidson, "A Coherence Theory of Truth and Knowledge," 314.

60. Davidson, "A Coherence Theory of Truth and Knowledge," 314.

61. Davidson, "A Coherence Theory of Truth and Knowledge," 309.

62. Davidson, "A Coherence Theory of Truth and Knowledge," 307, 310.

63. Davidson, "A Coherence Theory of Truth and Knowledge," 312.

64. Davidson, "A Coherence Theory of Truth and Knowledge," 307.

65. Rorty, "The World Well Lost," 13-14.

66. And yet apparently common. Crawford Elder goes so far as to assert that such silliness can be a definition for realism, and in a tone which indicates he thinks it rather uncontroversial: "Roughly, [realism] asserts that the beliefs held by any individual or any group about the world—about the world in general or about these or those components of the world—might massively be false, no matter how carefully researched." Crawford Elder, "Realism, Naturalism, and Culturally Generated Kinds," *The Philosophical Quarterly* 39, no. 157 (1989): 425-6.

67. Rorty, "The World Well Lost," 14.

68. Davidson, "A Coherence Theory of Truth and Knowledge," 310.

69. Rorty, "The World Well Lost," 15.
70. Davidson, "A Coherence Theory of Truth and Knowledge," 310.
71. McDowell, *Mind and World*, I, 6n.
72. Of course, once there exists a relatively stable cognitive architecture, the causal signature of a "perception" must matter to its eventual content. But the cause has been given such significance by the conceptual matrix, responsibility for the content must be attributed to that matrix. For more on this see the last sections of chapter 4.
73. It may perhaps be the case that certain "interpretations" of certain given sensory-causes are "hard-wired" parts of our biology; our recognition of (and reaction to) pain may be a prime candidate. An evolutionary epistemology would exploit just such examples as indicating just the interpretive limitations which a realism demands. Yet we must be careful not to connect our capacity for conceptualization too closely with natural, deterministic brain functions. "We would not be able to suppose that it is conceptual capacities that are in play in experience if the capacities were manifested only in experience, only in operations of receptivity. The capacities would not be recognizable as conceptual capacities at all unless they could also be exercised in active thinking, that is, in ways that do provide a good fit for the idea of spontaneity." McDowell, *Mind and World*, I, 5.

For further complaints against biophysical determinism and its jettisoning of thought see my Introduction to *The Incorporated Self: Interdisciplinary Perspectives on Embodiment* and Joseph Margolis, "A Biopsy of Recent Analytic Philosophy" *Philosophical Forum* 21, no. 3 (Spring 1995): 161-88. It seems unlikely that scientific progress would have taken the form detailed by Thomas Kuhn (for example) if knowing were a largely bio-physically determined phenomenon. Quine in particular is aware of this problem and posits a realm of freedom ("the ingenuity to rise above [natural selection]") to account for our capacity to theorize scientifically. (See Quine's "Natural Kinds" in *Naming, Necessity and Natural Kinds*, edited by Stephen P. Schwartz [Ithaca, NY: Cornell University Press, 1977.]) Such a concession threatens the entire nativist/physicalist project, for by placing cognitive activity in a free realm, it must also place perception there, or return to the simple skeptical empiricism which epistemic physicalism was meant to overthrow.
74. McDowell, *Mind and World*, I, 6.
75. McDowell, *Mind and World*, I, 6.
76. McDowell, *Mind and World*, I, 6.
77. McDowell, *Mind and World*, I, 4.
78. McDowell, *Mind and World*, I, 4.
79. McDowell, *Mind and World*, II, 2.
80. McDowell, *Mind and World*, II, 3.
81. McDowell, *Mind and World*, II, 3.
82. McDowell, *Mind and World*, VI, 1.
83. McDowell, *Mind and World*, VI, 1.
84. Cannot "in themselves" do any work for us. As I argue in more detail in chapter 4, once a relatively stable conceptual-linguistic architecture is in place to receive the deliverances of sensation, then the particular causal signature of a given sensory experience comes to have conceptual significance. Thus once we reliably interpret certain sorts of signals from rods and cones as "red," and certain as "green," then the particular causal

signature of these receptions (which we never experience as such) can serve as an indicator of the applicability of a given color-concept. The point is rather that the original nature of the causal impulses cannot justify any particular world-picture, and this will be so even if the causal signature were to undergo change, as with spectrum inversion glasses and the like. Although things might look "weird" for a bit, the conceptual-linguistic always supervenes on the causal.

85. Note that this needn't imply that the Categorical Intuition provides us originally and definitively with all the concepts we utilize. We may very well be limited in our understanding of the world by our current stock of concepts which are available to be drawn into operation, while at the same time Categorical Intuition is sufficient to allow the world to act as a guide as we modify, multiply and otherwise refine that set of concepts.

86. For an illuminating account of how Categorical Intuition can be brought to bear on just the sorts of problems raised in this work, see Richard Cobb-Stevens, *Husserl and Analytic Philosophy*.

87. The third dogma was identified by Davidson: "I want to urge that this . . . dualism of scheme and content, of organizing system and something waiting to be organized, cannot be made intelligible and defensible. It is itself a dogma of empiricism, the third dogma." "On the Very Idea of a Conceptual Scheme," in *Inquiries into Truth and Interpretation*, 189.

88. One can, of course, simply use the term "perceptual" differently from the way it is generally used, and insist for instance that McDowell maintains that our openness to the world is entirely perceptual, although he has redefined perception to include a non-causal sensitivity to the basic structure of the world. Such a claim would leave the point I am urging unaffected.

89. Davidson's conceptual operations go beyond those of our sense-organs, but are precisely not open to the world. Davidson cleverly argues against the possibility of *alternative* conceptual schemes—thus apparently concluding that there can be at most one account of reality, and that we cannot be radically mistaken—but because the only link between self and world is causal, his picture does not have room for the claim that the (objective) independent world influences the shape of this one account.

90. Part of my uncertainty about McDowell's ultimate epistemic position rests on the fact that late in the lectures he insists on the centrality of the fully embodied self to any understanding of our epistemic position in the world. (*Mind and World*, V, 5.) Further, his insistence on using the notion of a "second nature" (in the Aristotelian sense) to bridge the gaps opened by Davidson, Rorty, et. al., may well indicate that he takes the conceptual to be open to the world in virtue of practical activity (habits and the like). Unfortunately, he does not say explicitly what he means by the fully embodied self, or why exactly, it must be central to his account. Thus I am concerned that the notion of experience which McDowell inherits, and the account of embodiment which it implies, is still too close to the rationalist-empiricist (neo-Kantian) one to work for him in this context. Carlos Thiebault has suggested in conversation that, while he is sympathetic to my concern about McDowell's account of experience, still McDowell's account taken on its own terms is closer to my own than I seem willing to allow. I will be quite pleased if McDowell can find much to agree with in my own understanding of the epistemic import of the embodied presence of the self in the world.

4
Embodiment and the Epistemic Availability of the World

4.1: Locke's Gold: Nominal Essences and the Limits of Knowing

It may seem odd to begin an account of the availability of the world and its contents with Locke, since we are in various ways removed from his thinking; most importantly, we are at least three dogmas away from his empiricism. But this apparent philosophical distance actually serves my point, for what I hope to show in the next two sections is that the knowing subject in contemporary philosophy still stands in essentially the same relation to the (would be) known world as she did in Locke's philosophy. If I am right, this is a problem that needs attention, for it indicates continued allegiance to an epistemic position which has been shown to be in various ways untenable. I have been arguing that at the root of this position is a belief in a single, sensation-based mode of epistemic access to the world; that such otherwise different epistemic systems could display the same limitations (and tendencies to anti-realism) is (I would think) a strong motivation for reconsidering our allegiance to this particular epistemic tenet. Whatever doubt about this claim remains after the arguments of chapter 3, I expect to be erased by the end of this section.

But there is a more specific reason for writing on Locke. Although we have claimed that a realist epistemology depends on the possibility of attributing our knowledge to the world, it is not yet clear what this implies for a realist *semantics*, i.e., a realist theory of reference and truth. Locke is of particular interest because he allows reference to *individuals*, but not to sorts (that is, he is a semantic realist about individuals but not sorts). I will argue that, given Locke's epistemology this was the right position for him to hold; analyzing the differences in his treatment of individuals and sorts will, it is hoped, illuminate the epistemic requirements of semantic realism.

As is well known, ideas are the center of Locke's empiricism. His thoughts on the nature, limits and scope of our knowledge derive ultimately from his position on ideas. As Michael Ayers has noted, the empiricism which Locke advances was "essentially a concept-empiricism, according to which all our ideas ultimately derive from experience, rather than the stronger, or at least different, view, knowledge-empiricism, according to which all propositional knowledge is empirical, ultimately based on sensory knowledge."[1]

For Locke the primary building-blocks of our picture of the world are simple ideas, "which since the mind, as has been showed, can by no means make to itself, must necessarily be the product of things operating on the mind, in a natural way, and producing therein those perceptions which by the wisdom and will of our Maker they are ordained and adapted to."[2]

These simple ideas, which include such qualities as temperature, color, texture and weight, impress themselves on our mind through the action of the sense-organs; but in so doing they also bring with them a notion of their origin, thus: "It is therefore the *actual receiving* of ideas from without that gives us notice of the existence of other things, and makes us know, that something doth exist at that time without us, which causes that idea in us."[3]

At first this looks familiarly Cartesian; we realize that a certain set or series of simple ideas is appearing before our mind, and, through a process of reasoning, we infer that (having ruled ourselves out as the origin of this series of sensations) they are caused from without. Importantly, only if we can make this judgement can we count the experience as knowledge.[4] But Michael Ayers, quite rightly I think, is at pains to distance Locke's account from Descartes'. For Locke, the senses do not require the consent of judgement to count as a knowledge-producing faculty: "Sensitive knowledge is an immediate awareness of an idea as a certain sort of effect . . . the idea of white represents the quality of white conceived as an external power. Hence sensitive knowledge is knowledge of the existence of such qualities, whatever actual attributes give rise to them."[5]

For Locke, then, we have immediate, non-inferential knowledge of the existence of objects impinging on our senses. Interestingly, this knowledge is cast primarily in terms of our awareness of the cause-and-effect relations which obtain between us and these outside entities. Thus: "If anyone will examine himself concerning his notion of pure substance in general he will find he has no other idea of it at all, but only a supposition of he knows not what support of such qualities which are capable of producing simple ideas in us; which qualities are commonly called accidents."[6] And: "Powers therefore justly make a great part of our complex ideas of substances."[7]

Thus coming with our simple ideas is a notion of that (we know not what) from which those perceptions flowed, and this unknown something is associated for Locke with the real essence or constitution of the object responsible for our perceptions. Locke writes: "[N]atural things . . . have a real, but unknown constitution of their insensible parts; from which flow those sensible qualities which serve to distinguish them one from another."[8]

It is important to note that this account assumes our capacity to know individuals as such, as the "things" which posses the constitutions which affect our senses. For Locke, individuals could be known by experience alone (i.e., without the need for a mental synthesis of disparate sensations/representations) because the senses give us knowledge not just of the secondary qualities noted above, but also of those aspects of matter (primary qualities) which it must have in order to *be* such. Solidity, figure, extension and motion are the minimal properties which we must attribute to matter in order to have the basis for a materialist or corpuscularian explanatory system. Because Locke imagines the senses could carry this sort of information, there is for Locke's empiricism no individuation problem akin to the one which arose in the course of the investigations of chapter two; the problem of synthesis occurs for individuals only after we have discarded, as did Hume, the distinction between primary and secondary qualities.[9] For Locke, when we sense the solidity or extension of a thing, what we sense is present in the thing itself, and therefore we are assuredly presented with a substantial individual. No individual is literally "green" (rather, it literally possesses the power to cause green in us) but a thing *can* be literally spherical, incompressible, and at a given location.

But while accepting in principle the idea of a real essence to things, from which our simple ideas flow, and to which (beyond themselves), our perceptions point, Locke insists that our knowledge of this constitution is severely limited. Although our Maker has given us the faculties such that we might perceptually discern the appropriate actions to take in response to the world (so that it is proper to recoil from the excessively hot, eat the tasty, tread upon the solid,[10] and so on), and although on this basis we might safely surmise that the interactions of the real substance underlying the excessively hot with that underlying our bodies would produce negative results, while that between us and that "tasty" morsel would be beneficial, "we have no knowledge beyond that, much less of the internal constitution and true nature of things, being destitute of the faculties to attain it."[11]

Thus although Locke is quite clear in his contention that "it is the real constitution of [a thing] . . . on which depends all those properties of color, weight, fusibility, fixedness, &c., which are to be found in it,"[12] given that we can have no knowledge of this real constitution, we can never make the claim that any two individuals possess the same real essence. When it comes to our division of the world by genera, by the sorts "man," "gold," "horse," and "water" in which our thought so frequently traffics, Nature cannot herself be our guide. Instead, our complex ideas of genera, being what Locke calls "Nominal Essences", are built up by concatenation of simple ideas, thus:

I would not here be thought to forget, much less to deny, that Nature, in the production of things makes several of them alike. . . . But yet I think we may say, the sorting of them under names is the workmanship of the understanding, taking occasion, from the similitude it observes amongst them, to make abstract general ideas, and to get them up in the mind, with names annexed to them as patterns or forms.[13]

The genera of things, which we might through confusion or lack of reflection take to correspond to actually existing essences, are rather the work of the understanding, which through experience and practical interest unites those simple ideas it encounters into complex ideas, representing those qualities it most expects to find together in its future encounters with the objects of the world.

Now since nothing can be a man, or have a right to the name man, but what has a conformity to the abstract idea the name stands for (That is, has the attributes which have been chosen by us to make the connotation of the name 'man'.), nor anything be a man, or have a right to the species man, but what has the essence of that species: it follows, that the abstract idea for which the name stands, and the essence of the species, is one and the same. . . . It is true, there is ordinarily supposed a real constitution of the sorts of things; and it is past doubt there must be some real constitution, on which any collection of simple ideas co-existing must depend. But, it being evident that things are ranked under names into sorts or species, only as they agree to certain abstract ideas, to which we have annexed those names, the essence of each *genera* or sort, comes to be nothing but the abstract idea which . . . the name stands for.[14]

Our sorting of the world into kinds is a *de dicto* matter, "adjusted to the true ends of speech" and carries no connotation that real essences answer in nature to the abstract ideas of nominal kind terms. For real essences "come not within the reach of our knowledge when we think of those things, nor within the significance of our words when we discourse with others."[15] And given this limitation we can never know when something with a given *real* essence is before us. "For let it be ever so true, that all gold, i.e. all that has the real essence of gold, is fixed, what serves this for, whilst we know not, in this sense, *what is gold or is not gold?* For if we know not the real essence of gold it is impossible we should know what parcel of matter has that essence, and so whether *it* be true gold or not."[16]

The most we can know, when presented with some set of individual parcels of matter which affect us sensibly in a way concomitant with our complex idea of "gold," is that each of these parcels of matter have a real internal constitution suited to produce in me the very ideas which I take to be indicative of gold. It is a premise of Locke's thinking that the particular effects which individuals have on us is due to their real internal constitution, but because our knowledge of secondary qualities is limited precisely to these sensible effects, we can have no evidence for anything like a one-to-one correspondence between real essences and sets of secondary qualities.

We can, after all, easily imagine three distinct, unknown, internal constitutions nevertheless sufficiently similar so as to produce in us ideas which together are consonant with the nominal essence of gold; likewise, we can imagine that the same real constitution under different circumstances affects us differently, and so in those situations falls for us under respectively different nominal essences. Worse still, we can imagine that these three "gold"-types interact differently with (let us imagine) three types of *aqua regia*, while nevertheless *appearing* to interact identically. All nine possible gold-*aqua regia* combinations lead to the dissolution of the gold! Individually and in combination, it is easy to imagine these "substantially" distinct compounds all having the same sensible effects; causal/mechanistic difference may remain forever hidden.[17]

It nevertheless remains true that real essences are the true natural agents in Locke's ontology. It is in virtue of having a real, internal constitution that any given individual has the power to affect others the way it does. This is in keeping with the materialist, corpuscular philosophy which he takes from Newton, Boyle and others. But what is interesting about this position is that, since for Locke substances are the interacting agents of Nature, science must continue to take an interest in them; yet it is the very epistemic relation which we have to substances, which even while providing the foundation for practical affairs, must block our attempts at true Natural Philosophy. We usefully define and group the individuals of the world in terms of their varying powers to affect our senses; but for science to be possible we must know not only the mechanism of these effects, but also the causal relations of substances to one another.

> We are so far from being admitted to the secrets of Nature . . . for we are wont to consider the substances we meet with, each of them, as an entire thing by itself, having all its qualities in itself, and independent of other things; overlooking, for the most part, the operations of those invisible fluids they are encompassed with, and upon whose motions and operations depend the greatest part of those qualities which are taken notice of in them, and made by us the inherent marks of distinction whereby we know and denominate them.[18]

Although even this much knowledge would be for Locke no better than speculation (albeit reasonable, scientific speculation), it is clear that he thinks that Nature sorts things according to shared internal constitutions, *which are to be understood in terms of their causal powers and tendencies*, that is, precisely in terms of their relations to everything else.

> If inanimate bodies owe so much of their state to other bodies without them, that they would not be what they appear to us were those bodies that environ them removed; it is yet more so in vegetables, which are nourished, grow, and produce leaves, flowers and seeds, in a constant succession . . . And if we look a little nearer into the state of animals, we shall find that their dependence, as to life,

motion, and the most considerable qualities to be observed in them, is so wholly
on extrinsical causes and qualities of other bodies that make no part of them, that
they cannot subsist a moment without them . . . Take the air but for a minute
from the greatest part of living creatures and they presently lose sense, life and
motion. This the necessity of breathing has forced into our knowledge.[19]

Locke's analysis carries both epistemological and metaphysical import.
Metaphysically, individuals are sorted into kinds in virtue of substantial similarity,
simultaneously defined in terms of causal relations to other substances and internal
corpuscular structure. But we can have no epistemic access to this sorting.
Rather, epistemic sorts are concatenations or syntheses of simple properties. The
synthesis can be guided by experience, practical needs and interests, scientific
hypotheses or the standards of global taxonomic systems; there may be good
reasons for sorting the world the way we do, and there may be criticism and
change of our current set of nominal essences. In all this we are no doubt guided
by the natural light of Reason. Locke's worry is not that our choices are arbitrary,
but that because of our epistemic limitations our choices are always guided by the
phenomenological order, by the structure of the *appearance* of the world, and not
by the natural causal order itself. Although as previously noted we can be assured
that the phenomenological order is largely appropriate to the causal, i.e., that it
provides a fine practical guide, it does not follow from this assurance that the real
causal relations of things, which make the world operate as it does, are thereby
revealed.[20]

> Before we can have any tolerable knowledge of this kind, we must First know
> what changes the primary qualities of one body do regularly produce in the
> primary qualities of another, and how. Secondly, we must know what primary
> qualities of any body produce certain sensations or ideas in us. This is in truth
> no less than to know all the effects of matter, under its divers modifications of
> bulk, figure, cohesion of parts, motion and rest, which, I think, everybody will
> allow, is utterly impossible to be known by us without revelation.[21]

The epistemic problem related here, which drives the whole of Locke's
inquiry, is not rooted in the vast complexity of nature, which perhaps exceeds the
capacity of the human mind to grasp. For it is an assumption of the materialist
philosophy that there is a design to the universe, based on ultimately similar bits
of "matter in motion," placed in differing configurations, and in principle such a
non-random universe is definable by a finite set of mechanistic laws. Instead,
what Locke says is that it would be impossible for us to know the causal laws of
the universe *without revelation*, i.e., without having a method of *gathering*
knowledge more direct than sensation. It is because Locke is convinced of our
limitation to a single, sensual mode of epistemic access to the world[22] that he
despairs of the possibility of a true empirical Science of Nature: "This way of

getting and improving our knowledge in substances only by experience and history . . . makes me suspect that Natural Philosophy is not capable of being made a science. We are able, I imagine, to reach very little general knowledge concerning the species of bodies, and their several properties."[23]

For this reason, the metaphysical and epistemic analyses of Locke's *Essay* are both complimentary and divergent: we are involved in the causal order, indeed, it is in virtue of our causal contact with substances that we know them at all. And for Locke this causal contact indeed results in knowledge; because primary qualities are aspects of things themselves, the causal conduit which brings us our ideas of material particulars as extended, placed, solids gives us to this degree real access to the physical structure of the universe.[24] We are not for Locke trapped by the phenomenological deliverances of our senses in a way which implies epistemic distance from the particulars which produce them.[25] But the same cannot be said for our knowledge of kinds. Although the reception of secondary qualities gives us knowledge of the real powers of objects, this knowledge is only of the power to cause in us precisely the ideas in terms of which our knowledge of the powers is cast. This severe limitation of our epistemic access drives a wedge between us and the world with respect to our genera, making divergent the epistemic and metaphysical tenets of Locke's philosophy which are for material particulars complementary.

Locke's distinction between primary and secondary qualities, and thus his different position on our knowledge of material particulars and on genera, can be seen as a distinction in the degree to which the subjective world of appearance becomes the object of our knowledge. When we know the particular, we know it in terms of space, place and extension which, even if for Locke they are presented to us by sensory experience, are not in themselves defined by the quality of that experience, nor is the knowledge thereby gathered *cast* in terms of the sensory experience: what is known is *that the world is this way*.[26] Thus, the object or referent of our idea is the structural arrangement of the world itself. By contrast, in our experience of secondary qualities, the experience is defined precisely in terms of its subjective quality; what is known is *that I am experiencing such-and-such*. Even my world-indicating knowledge *that the world has such-and-such power* is defined and delimited in terms of the subjective experience (the power to produce in me this experience).[27] Thus, even though our *reception* of primary and secondary qualities is formally identical, involving the direct, non-referential awareness of the character of the world, their epistemic import is distinguished by the different status of the foremost *objects* of the knowledge thereby gained.[28] As has been outlined already, the concatenation of secondary qualities into sorts is the work of the understanding, operating on knowledge immediately received as simple ideas of secondary qualities, and so regardless of the care with which the sorts are constructed the object of our knowledge can only be the complex ideas themselves. Nominal essences, this is to say, are entirely artifacts of the phenomenal order, and while the phenomenal order has the connections with the causal order already mentioned, divisions of (made in) the phenomenal order needn't correspond to any such divisions of the causal. The world does not impress nominal essence upon us as it does simple qualities; our ideas of genera (the particular concatenations) are not, and cannot be directly attributed to, the world or its effect on us.[29]

It is for this reason that Locke insists that we cannot refer to kinds (or, rather that in taking about kinds we refer to our ideas and no further, for real essence "comes not . . . within the signification of our words"[30]). For Locke's empiricism provides the resources (our reception of primary qualities) whereby we can have an epistemic link with individuals (and can therefore refer to them), but we can have no similar link to sorts; in one case, but not the other, our *knowledge* reaches the world. This indicates that knowledge and reference are deeply tied for Locke, and I think we will find this a connection worth exploring in greater detail. Looking at the issue through Lockean eyes may well help us see better our way out of current dilemmas.

4.11 Locke on Ideas, Intentions, and Reference

I have argued above that the reason for Locke's insistence that we cannot refer to natural kinds is that he thinks that we can have no knowledge of natural kinds; for Locke the phenomenal order within which our nominal essences are defined is *epistemically closed*, in that these complex ideas cannot be epistemically attributed to the world. But for certain theories of reference this epistemic closure offers no hurdle to the possibility of reference. So I shall spend the next few paragraphs showing why Locke was precluded from accepting the most likely candidates, and also why *we* shouldn't accept them. Only then will I move on to outline the theory of reference which I think best illuminates Locke's intentions.

The most obvious candidate for a theory of reference suited to Locke's epistemology is the familiar "traditional" theory, (so dubbed by Stephen Schwartz in his introduction to what is perhaps the most definitive collection of criticisms of this model[31]) according to which the contents of the ideas associated with given terms dictate the extension of that term through some kind of "resemblance" relation. In the context of this sort of theory, the easiest way to explain Locke's insistence that we cannot refer to sorts is by analyzing it in terms of his claim that nominal essences do not exist in the world; we cannot, of course, refer to non-existent things. But this leaves open some problematic possibilities: If "proper names" and genera terms have the same referential *capacities* (refer by the same mechanism), but genera always fail simply because the world failed to contain anything corresponding to them, then this leaves open the possibility that, if our ontological suppositions were especially fortuitous, genera words might sometimes succeed in referring. Given Locke's insistence that there (probably) *are* natural kinds, a real possibility of such correspondence arises. But such a conclusion would be out of step with Locke's intentions. Even if we were so lucky that the set of all individuals capable of producing in us a set of simple ideas corresponding to our nominal essence, was in fact co-extensive with a set sharing some real essence *in virtue of which* its members produced the ideas contained in the nominal essence of gold, it would not follow that "gold" succeeded in referring. (And it would certainly not follow that we could *know* that we had so succeeded.) It seems unpromising to analyze the different referential success of given words in terms of the likely ontological status of their intended object.

But given the outcome of this investigation, another possibility may well arise: if reference proceeds along lines of phenomenological resemblance, we could interpret this resemblance relation more strictly, allowing it to explain our capacity to refer to individuals via primary qualities (for they *actually* possess a "resemblance" to the ideas they produce); whereas genera terms would be precluded from referring since the real essence in terms of which a given natural kind may be defined would not, strictly speaking, "resemble" the secondary qualities making up the nominal essence. But as this analysis would make secondary qualities referentially irrelevant, it would lead to a more generally unsatisfactory theory of reference. For it seems our capacity to refer to *particular* individuals can at least in some cases be attributed to our possession of an identifying phenomenal description, e.g., "the blue one."[32] According to a theory of reference operating according to strict phenomenological resemblance, since no aspect of the blue object would resemble my idea of blue, the description would not aid or guide my reference. In saying "grab the blue thing off the table," I may well have referred to a thing on the table (for I can have access to particulars and their spatial locations in terms of primary qualities), but I would not have referred to any particular individual so placed because my description, failing to resemble any given particular, would not have allowed for definitive reference. But it would be foolish to deny that visual-descriptive discrimination can play an important role in differentiating objects, and is likely, therefore, to play some role in the possibility of accurate reference.

What *is* clear, however, is that phenomenological resemblance relations represent a bad way to analyze reference, even if we loosen the "resemblance" requirement, allowing secondary qualities to guide reference in virtue of their ability to produce in us ideas which resemble the ideas which guided the reference. It is simply much too easy to construct counter examples entailing mistaken reference, thus: In the case of primary qualities, we might imagine a given set of ideas constituting our notion of a given particular (five inch diameter, five gram spheroid in the kitchen) in fact better resembles an object whose existence we did not suspect (say someone just placed it in the kitchen) than it does the object we intend. Unbeknownst to us, and without our consent, we have referred to the wrong ball! The counter examples are even easier and more obvious with secondary qualities: I may refer to "the man with champagne in his glass," only to discover that he was drinking ginger ale. On a theory according to which the *content* of my idea or description determines reference, I would have referred to someone else.[33] But I needn't give up on the notion that my description played a role in helping *me* identify or keep track of the guy to deny that the inaccuracy of that description implied errant reference. Likewise, if I have on color-altering contact lenses and, looking at an array of objects I ask for "the blue one" (which is in fact green), I have *not* thereby referred to the one which is in fact blue (however it looks), even though it better matches the qualitative content of my "image."[34] These considerations throw into serious question the idea that words refer in virtue of the content of their associated ideas.[35] Indeed, to suppose that simple ideas could guide reference in virtue of their content would re-open the possibility that *groups* of secondary qualities, and therefore nominal essences, could refer. And if we wish to reproduce Locke's intentions, this is a door we must leave closed.

It would, I think, be much better to understand Locke's position here by reference to the concept of an epistemic link: In the case of direct reference, for instance, I know (or mean) *that one* because that is the one which is affecting me, which is producing in me the "simple ideas" (whatever their content), in terms of which I may well articulate what I take to be identifying descriptions, and which may even operate as such for me.[36] And in the general case, I think it is fair to say that, for Locke, our capacities for, and the limits of, reference correspond exactly to the capacities and limits of our epistemic access to the world. I can refer to individuals, as such, and to powers, as such, *in the world*, because I can have epistemic access to these features of the world.[37]

In the case of nominal essences, in the context of Locke's epistemology, it is not possible to refer because it is not possible to have an epistemic link with the natural kind (the "real essence" or that in virtue of which the members of a kind are unified) which such a complex idea might ostensibly represent. Having an epistemic link only to "powers" or secondary qualities is not equivalent to having an epistemic link to that in virtue of which a given individual is a member of a natural kind; thus the epistemic link which can account for our reception of simple ideas cannot also account for our formation of complex ones. Alternately, we can say that there is, within the context of Locke's epistemology, no epistemic link to anything to which the complex idea could be attributed:[38] *we* produce the complex idea, so the epistemic link terminates, along with the reference, in our own ideas. Even in the case of that fortuitous circumstance whereby the extension of a given nominal essence (gold) matched that of a given natural kind, it would not follow that "gold" picked out some real thing in the world; for without an epistemic link to the essence, responsible for our complex idea (to which our complex idea could be attributed), we would not have *picked out* anything in the world by forming the idea. Without the epistemic links in virtue of which an *attribution to the world* is possible, *reference to the world* also fails.

The complex phenomenal order—the contents and relations of complex ideas
—is for Locke, in this sense, epistemically closed. Our knowledge does not reach
beyond the phenomenal, and for this reason neither do our words. Even simple
ideas of secondary qualities, while causally open, are epistemically closed to the
degree to which our knowledge is primarily *of* the phenomenal qualities
themselves. Of course, and this is important for Locke, as it is for certain
contemporary theorists, along with these simple ideas comes knowledge of their
origin in the world; but we always have to remember for Locke that if we are in
position to receive information about secondary qualities, we have, in virtue of our
reception of primary qualities, already individuated a particular from which that
information came. According to Locke's peculiar empiricism I have an epistemic
link with the world sufficient to distinguish particulars, and therefore sufficient to
support the determination of particular objects of reference. Without the epistemic
access afforded by our reception of primary qualities, knowledge of *causal* origin
would go no further than the awareness that the ideas were caused *in* us, and not
by us. *What* they were caused by, or what relation that information bore to the
structure of reality, would remain forever beyond the veil of sensation.[39] Thus,
more to the point, insofar as this degree of epistemic connection would not enable
us to distinguish individuals, we also lose the ability to *refer* to individuals.
Epistemic incapacity implies referential incapacity.

4.12 Putnam's Water: H₂O as Nominal Essence
Hilary Putnam is one contemporary thinker who embraces this conclusion.
Indeed, I take it to be an obvious fact—in need, perhaps, of elaboration but not
argument—that he is an "internalist" or "pragmatic realist" for whom what we
have been calling the phenomenal order is epistemically closed.
   Usefully, we can begin from Putnam's analysis of what follows from Hume's
critique (and Kant's acceptance of that critique) of Locke:

If *all properties are secondary*, what follows? It follows that everything we say
about an object is of the form: it is such to affect *us* in such-and-such a way.
Nothing at all we say about an object describes the object as it is 'in itself' . . .
[But] 'All properties are secondary (*all properties are Powers*)' suggests that
saying of a chair that it is made of pine, or whatever, is attributing a Power (the
disposition to appear to be made of pine to us) to a noumenal object; saying of
the chair that it is brown is attributing a different Power to that *same* noumenal
object; and so on. . . . [And] Kant [who Putnam is here using to argue for
internalism] explicitly *denies* this. . . . On Kant's view, any judgement about
external or internal objects . . . says that the noumenal world as a whole is such
that this is the description that a rational being . . . given the information
available to a being with our sense organs . . . would construct. In *that* sense, the
judgement ascribes a Power. But the Power is ascribed to *the whole noumenal
world*; you must not think that because there are chairs and horses and sensations

in our representation, that there are correspondingly noumenal chairs and noumenal horses and noumenal sensations. *There is not even a one-to-one correspondence between things-for-us and things in themselves.*[40]

Once we deny Locke's distinction between primary and secondary qualities (or, more precisely, deny that we have any epistemic warrant to maintain that distinction) then our ideas of individuals, no less than our ideas of sorts, become nominal essences. For Putnam, individuals and kinds are divisions of the phenomenal order, made by different, complementary, sometimes competing criteria including simple qualities (color), practical interests (hunger), and scientific and quasi-scientific theories (as those governing beeches). We "make" objects by imposing (whether voluntarily or not, as a condition of experience, or *a posteriori*) descriptive schemes on experience: "'Objects' do not exist independently of conceptual schemes. *We* cut up the world into objects when we introduce one or another scheme of description."[41] What makes Putnam's account so attractive is his liberalism with regard to what can be a legitimate factor in determining our collective ontology (and our ontology *is* collective; this is very far from a solipsism). Indeed, I find it useful to understand Putnam's views in terms of a kind of pragmatic-phenomenal ecology, whereby each "individual" and "kind" is a node in an interlocking web of relations understood and dictated by theories, interests, etc., themselves bound up with, dictated by, and relative to our various epistemic (political, cultural, scientific) communities, which ecology is perceptually effective (so that we see the world in terms of this socio-linguistically instantiated[42] ecology). Our knowledge is in some sense a product of causally operating receptivity, although this realization has little epistemic value, as our interpretations of these causes are pre-consciously dictated by said perceptually-active ecology. On this model, cause is something like the electricity which powers the knowledge producing machine: it doesn't dictate which images or descriptive schema are produced (that is a product of the conceptual structure of the mechanism) but without the input there would be no images at all. As Putnam puts it, "internalism does not deny that there are external *inputs* to knowledge . . . [but] it does deny that there are any inputs *which are not themselves to some extent shaped by our concepts* . . . or any inputs *which admit of only one description, independent of all conceptual choices.*"[43]

The structure of reality is not available as such just in virtue of our causal connection with that reality. As we saw in chapter 3, as Locke clearly saw, and as Putnam here argues, an epistemically closed internalism, idealism or conceptualism is a straightforward consequence of any epistemic model which inserts judgement or synthesis between our reception of sensory input and our apprehension of perceptual knowledge.

But what is it from which one is thereby epistemically (and, I have argued, semantically) distanced? Let us adopt a brief terminological convention to help with this question: a contrast between the older sense of "object"—that towards which the consciousness is directed, anything present to mind—and the solid, physical ("real") "thing." We have seen in various ways that to approach things (and kinds) as by definition the bearers of perceptually available properties (where perception was thought to exhaust our epistemic resources) opens a gap between the epistemically available qualities of the thing (which form for us the "object"), and its metaphysically vague *thingness*.[44] So lest we make all things real and unknowable, or available and artificial, we should reconsider the nature of the things to which we desire access.

For it is clear that when an individual appears to us as *object*—as perceptually-based representation—it thereby fails to be for us a thing, and fails in just the ways which give rise to the skeptical attitude. For in this context, where the object (the representation) and the sensible characteristics which constitute it are meant to constitute our knowledge of the thing, we have a right to ask about the accuracy of the mode of access, and the appropriateness of the principle guiding the synthesis, which together constitute our apparent knowledge. The object is present to us as *individual*, yet in terms of a matrix and medium (sensible qualities) which cannot itself account for our capacity to *individuate*.[45] We must ask why just these qualities should belong to the object? Why should it have just these apparent limits? Given only one mode of epistemic access, the obvious (if optimistic) reply that the boundaries of the object are such precisely because they correspond to the boundaries of the thing is precluded; the boundaries of the *thing* are not perceptually, and hence not epistemically, available to us. The principle of division remains mysterious, hidden and suspect.

Yet to do justice to epistemic (and metaphysical) realism, for our knowledge of the boundaries of objects to be sufficiently robust to support truth, we need to be able to make sense of a relation obtaining between the object and the thing. Insofar as the boundaries of the object are just that, boundaries which belong to the object and its representational matrix, what is missing is the sense that these boundaries belong also to the thing. There is no justification for the attribution to the thing of the boundaries which the object represents it as having. This epistemic picture does not allow the thing to possess, to be responsible for, those boundaries. When epistemology is sensation-based, then, it becomes natural to propose the existence of an outside world, the individuation of which into objects nevertheless depends upon us.[46] What remains disastrously hidden as long as the thing, as object, exists for us only as object-representation, is the fact that as thing it exists for itself as an independent, self-standing[47] particular.[48]

The metaphysics of thinghood which we must satisfy, then, are not outlandish or taxing: while allowing there to be a relation between object and thing, we must preserve the sense of the self-standingness of the thing. Our epistemic access to the thing and its boundaries must recognize its metaphysical independence as such, and, as in chapter 3, we must be able to attribute *to* the thing our knowledge of its boundaries. But Putnam, with the mainstream of contemporary epistemology, has accepted an empiricism which, based as it is on a conceptual synthesis of perceptually available properties, is necessarily epistemically closed. Thus we *cannot* know which or what of our divisions corresponds to divisions in the "real" world.

Direct Realism may offer a way out. According to Direct Realism, reference is determined not by the epistemic state of the language user, but by the world itself. This seems to promise a degree of access to the "thing itself" (whether this "thing" is a particular or sort), sufficient to attribute our reference to *divisions* of the world *to* the world. But in this *epistemic* context the assertion that "the world determines reference" does no work at all. It is a purely formal stipulation that our words refer to something, some "real essence" in the world, because our knowledge is occasioned in the global sense by causal links. Putnam himself chides this sort of direct "metaphysical" realism for imagining that words can somehow refer apart from our conceptions of the world.

> The metaphysical realist formulation of the problem once again makes it seem as if there are to begin with all these objects in themselves, and then I get some kind of lasso over a few of these objects ([those] with which I have a "real" connection, via a "causal chain of the appropriate kind"), and then I have the problem of getting my word . . . to cover not only the [objects] I have "lassooed" but also the ones I can't lassoo, because they are too far away in space, or time, or whatever. And the "solution" to this pseudo-problem . . . the metaphysical realist "solution". . . is to say that the word automatically covers not just the objects I lassooed, but also the objects which are *of the same kind*—of the same kind *in themselves*.[49]

The trouble is that, *ex hypothesi*, we do not know this kind-determining real essence in virtue of our reference-preserving connection to it, any more than we refer to it in virtue of our knowledge. Thus, in the context of any Putnamesque epistemic internalism (which follows neatly from the premises with which we have been dealing), we could never know *if* there were an individual or kind to which we referred, nor could we ever know if or when our knowledge of the individual or kind matched the actual referent of our words. In the absence of particularity in my *epistemic* link to the world, insisting on particularity in my *referential* access looks unmotivated. In fact, it borders on incoherent, and certainly approaches the philosophically silly, to assert that one can refer *without knowledge of one's referent being even theoretically possible*.[50] A doctrinaire Direct Realism offers no viable alternative to the possibilities sketched above. What was so pernicious about the "descriptive idea" model or "traditional" theory of reference is that it supposed that words, in virtue of their associated ideas, did things on their own, independent of the intentions of—and even without the knowledge and out of the control of—the language user. This sort of perniciousness is not avoided by a theory which takes reference to be a function of some non-epistemic relation between a language user and the real essence of an individual or kind.

Putnam, of course, is precisely *not* this sort of doctrinaire Direct Realist. For although he apparently accepts the claim that "the world determines reference," this does not, for Putnam, imply that reference proceeds independently of the epistemic and intentional state of the speaker. The reason, as any reader of Putnam will know, is that for Putnam the world *itself* is relative to the global conceptual-epistemic system of a given epistemic community: in his now famous phrase, "the mind and the world jointly make up the mind and the world." Putnam's acceptance of direct "internal" realism is based on the premise that the categories available as referents are defined by our global descriptive schemes:

> it is trivial to say what any word refers to within the language the word belongs to, by using the word itself. What does "rabbit" refer to? Why, to rabbits, of course! . . . When we use the word "horse" we refer not only to the horses we have a real connection to, but also to all other things *of the same kind*. At this point, however, we must observe that "of the same kind" makes no sense apart from a categorical system which says what properties do and what properties do not count as similarities . . . What makes horses with which I have not interacted "of the same kind" as horses with which I *have* interacted is the fact that the former as well as the latter are *horses*.[51]

The central point is that the extension of this term, although it is not defined by the mental state of any individual speaker, is nevertheless a function of global conceptual and epistemic considerations. For Putnam, as for Locke, reference is in various ways dependent on, and certainly cannot exceed, the epistemic powers of the individual (or that of the relevant members of his linguistic-epistemic community[52]). Putnam and Locke differ metaphysically, and therefore differ about which terms in fact refer, only because they differ epistemically. The general principle, that reference cannot exceed the bounds of an epistemically closed system,[53] is held in common. For each of them, to make sense of our capacity to refer we must suppose that capacity to be part of an integrated conceptual-linguistic mechanism relating us epistemically to the world. To suppose instead that reference is guided and determined by factors independent of our intentions and the epistemic capacities which support them (so that I am forced to imagine the possibility that the object I would pick out myself—pointing out the guy with champagne—diverges from the referent my words or 'ideas' would pick out) is to remove our capacity to refer from the only context within which it makes sense.

I began this section with two goals: strengthening the general argument of chapter 3 regarding the inevitability of the dilemmas generated by the fourth dogma of empiricism, and illuminating some general epistemic conditions for semantic realism. As for the first, I can say in my favor that both Putnam and Locke recognize the problem (and, it seems to me, compellingly lay out the available moves): Faced with the possibility of skepticism as a consequence of simultaneously accepting a "metaphysical realist" outlook and an epistemology rooted in sensation requiring conceptual synthesis, Putnam's "way out" is an historicized, intersubjective (not to say transcendental) idealism. His individuals and kinds are referentially and epistemically available, but this availability is purchased at the expense of making us the authors of the structure to which we refer: To coin a slogan, we might say that within the constraints imposed by the fourth dogma of empiricism, *available implies artificial*. For Putnam, H_2O is simultaneously nominal *and* "natural" kind.

Locke, too, faced with the necessity of "synthesis" in our formation of complex ideas reluctantly accepts an idealism of kinds. (To restate the slogan: *mentalism implies idealism*.) And Locke's stricter empiricism for individuals is far from secure. By insisting that experience itself can be a carrier of epistemic content, Locke invites Hume's probing with respect to the reliability of that content. *Empiricism implies skepticism.*

Given these theoretical binds, it seems that the second task of this chapter, to tie more firmly together epistemic and semantic realism, is a dangerous strategy. For if, as I have argued, *epistemic* access delimits referential access (and, more specifically, referential particularity is possible if and only if particularity in epistemic links is) then the above arguments about epistemology also threaten semantics.

We face the question: on what epistemic grounds can we pair epistemic (and therefore referential) availability with metaphysical reality? Answering this question is, of course, the task of the following chapters, but we can see already two *sorts* of options available to us. The first option is to follow a more thoroughly Lockean path, which after all purported to give us access to particulars, by maintaining that our epistemic contact with the world is entirely perceptual. The task facing the Lockean is to avoid Hume's critique, which *may* be accomplished by refusing to analyze our perceptual access to the world in terms of sensible qualia. Such a path is outlined by Michael Ayers in his masterful *Locke*, but it is not the path I will follow here. For me, part of the moral of chapter 3 is that traditional empiricisms offer no better support for realism than do the post-Humean ones.[54] I am committed to the accuracy of the analyses of sensual experience given by Quine, Davidson and others, and thus I am committed to their account of the status of perceptual knowledge. Thus the task I face is quite different: Against Putnam, I must make the case that the phenomenal order is *not* epistemically closed, and this, I argue, can be done by questioning the fourth dogma of empiricism, that sensation is our only mode of epistemic access to the world.

4.2: Space, Place, and Information: Realism and Material Particulars

As is by now obvious, preserving the viability of, and suggesting an epistemic framework for realism in metaphysics and semantics is a central concern which guides both the criticisms and the positive program of this work. By the end of our investigation of Frege, we came to see that direct reference, and with it our ability to think about material particulars, required not just a capacity to gather sensible information from the world sufficient to *identify* given particulars, but also an ability to *individuate* those particulars. And we have seen in various ways that to cast our epistemic access to the world entirely in sensible-causal terms jeopardizes realism, because it both threatens the notion that thought bears on reality and encourages the conflation of an object's criteria of sensible identification with its criteria of identity.

Thus, an acceptable account—which is to say, for the duration of this work, a realist account—of our ability to individuate material particulars, must include an account of our epistemic openness to the world which is not limited to sensation, which allows the metaphysical distinction between identification and identity to be maintained without inviting skepticism, and which accounts for the world's capacity to guide and direct our conceptualizations of it. I will argue that our spatial organization of perceptual information, and in particular our knowledge of the boundaries of material objects, depends upon both the comportmental *potentials* of the body, and its *actual* activity in and interference with the physical world. This line of thinking will lead to the conclusion that the linguistic-conceptual matrix which structures sensation so as to allow for the perceptual identification of particulars is itself both connected to the world causally (in virtue of perception) and open to the world via the body's activity.[55]

A similar line will be followed for our epistemic access to sorts; but here our fallibility becomes more of a factor to the degree that sorts are often the result of higher level (scientific) attempts to understand the world, rather than, as with a basic ontology of particulars, the precondition for such attempts. Thus the analysis I give (beginning in section 4.6) will aim primarily to provide an account of epistemic access which can support a convergent realism.

4.21 Russell's Principle

For the former argument, to which I now turn, I will be resting heavily on the analyses of Gareth Evans, and the ideal place to start is with his defense of Russell's Principle, on the acceptability of which, in a way, this whole project rests. Russell's Principle, an expression of a necessary condition for thinking about objects, can be stated quite simply: one cannot make a judgement about something unless she knows which something the judgement is about. The difficulty, as Evans notes, is not defending the principle so put, but spelling out what such knowledge amounts to.[56] Evans proposes to make the principle more definitive with the suggestion that what is required is "discriminating knowledge: the subject must have the capacity to distinguish the object from all other things."[57] This seems correct to me, although given my terminology here I should like to call such knowledge *individuating knowledge*.[58] It is clear that what Evans has in mind assumes what I would like to bring out with the term, and it can be illustrated with the following example:

Let us imagine a subject in possession of knowledge with content sufficient to identify one and only one thing, i.e., the thing in question is the only blue thing in the universe. Such knowledge has a claim to be discriminating knowledge insofar as it applies to one thing and no other. Now let us imagine a subject with the perceptual capacities necessary to detect blue and, when this occurs, to say "I see blue." Will the subject under consideration have the capacity to make the judgement "the *thing* is blue"? Evans' suggestion is that he will if and only if he has more than the capacity to detect blue (thus more than the capacity to utilize this sort of discriminating knowledge): in particular he must be able to differentiate the thing from its surroundings, what I should like to call the capacity to *individuate* the thing. Insofar as this goes beyond the capacities so far ascribed to our subject it involves at least the ability to know the extent of the blue thing, where its boundaries are. It involves the capacity not just to detect attributes as such, but to be able to ascribe all the attributes to things and to know when detected attributes belong to the same, and when different, things. For in effect the only capacity here displayed is the perceptual capacity to discriminate "blue" in one's perceptual field, and to attribute that perception to an external (objective) origin; but this capacity is not sufficient as such to identify any *particular* thing as the origin of the perception (and thus, despite the accuracy of the experience, such discriminating knowledge would not even be sufficient to support reference to the thing).

None of this is to deny the obvious: part of what would be required to be in a position to make judgements about a thing, to assert, or at least assent to, statements made about it, is to be in a position in which the information necessary to those judgements could be gathered. That is, one should have a certain epistemic sensitivity to the world, be in a position to detect attributes and qualities. But what differentiates an information link with a *particular thing* from the general capacity to gather information? Part of an answer can be given by considering a simple example: arrayed before me is a keyboard with at least one symbol on each key. What would differentiate the general capacity to gather information from the capacity to gather information *about individuals* would in this case be expressed by the ability, given the general capacity to know one is being presented with symbols, to know whether the "T" and the "D" were on the same key (they are not), and whether the "&" and the "7" were (they are). The ability to assert such truths would be a specific example of my general capacity to think and make judgements about keys, and thus an example of my capacity to individuate things.

From the point of view of the descriptivist, according to whom our knowledge of individuals is rooted in the sensory field, it would be natural to analyze my knowledge of (and capacity to individuate) some particular thing a in terms of the perceptually available attributes of a. My epistemic sensitivity to a would be explained in terms of knowing some set of a's attributes, some descriptions of a which may be sufficient to differentiate a from any given b. But it is clear that knowledge of descriptions is a poor way to analyze my epistemic capacity to individuate a particular. First of all, we need to be able to make sense of the possibility of thinking about (having individuated) a particular when one does not possess discriminating perceptual information; we can easily imagine a person having a cognitive or epistemic fix on one of a number of descriptively identical objects arrayed in a room.[59] And even in the case of nonidentical objects we rarely possess detailed enough information of a thing to plausibly claim that our capacity to individuate it rested on the accuracy and completeness of that information.

Second, to give an account of one's epistemic sensitivity to a particular (that which allows me to judge "this key bears the '&'") which centrally involves knowledge of *other* attributes of that particular is simply to put off the question at hand: how do we know that the attribute in question belongs to a particular thing? Epistemic sensitivity to a particular goes beyond the ability to detect specific occurrent properties insofar as it must involve differentiating between perceptual information pertaining to (concerning attributes of) some particular object rather than another. To put this another way: for any bit of information F, knowledge that F (detection of an F-attribute) will only support the thought a is F in the case that the subject knows that F pertains to (is about) a. Nor can it be said simply that assigning F to a is based on prior knowledge that G pertains to a, for although there are surely cases which fit this pattern, the same question arises about how we assigned G to a. Insofar as the required knowledge goes *beyond* knowledge of F or G, then the capacity to assign F to a cannot be based on knowledge of F. The ability to differentiate information pertaining to some a from that pertaining to some b (which ability constitutes the *particularity* of one's information-link/epistemic sensitivity to a given thing) cannot be based on the *content* of that information.[60]

The question we are facing, of course, is how, in attributing a given set of attributes and qualities to a given thing, we form the *object* by and in terms of which the thing is sensibly present to us; what we want to uncover is the principle of division for our sensory field. The sort of solution which this question invites is one which provides non-content based sorting criteria for any F sufficient to determine to which object a F is informationally related, and in so doing provides the basis for any concept of (the identity of) a which allows us to arrange our perceptions of a.

My suggestion is that these criteria can be understood in terms of the physical/spatial boundaries of objects, and that our knowledge of these boundaries can be accounted for by positing a mode of epistemic access to the world provided by the active body.

4.22: Ego-centered Space and Bodily Activity

To search for the criteria of identity for an object[61] is to seek that which will enable one to know, in each epistemically available case, what counts as (a part of) the object and what does not. For information about an object must pertain either to the whole or to some part; recognition that the information indicates some attribute of the whole ("spherical," say, or "red") implies (prior or simultaneous) knowledge of the extent or nature of the object as a whole. Likewise, to recognize that some bit of information pertains to a *part* of an object, indicates some sense of that object which goes beyond knowledge of the part in question. Still, the simple interpretation of sensory information (there's some blue!) can occur without clear sense of the identity or boundaries of the object from which the information came. So the question becomes: in virtue of what are we able to make the leap from interpretation of sensory information to the recognition of the attributes of objects?

We can get some distance toward an answer to this question by considering one obvious facet of our capacity to recognize objects and their parts: for information about a part of an object to be useful, that information must indicate not simply which object it pertains to, but which *part* of the object it pertains to; to identify a part of an object is also necessarily to identify a place on an object, a place from which the information derives. Information is not merely sorted according to object-of-origin, but it must also be structured according to place-of-origin. There is other evidence for this close relation of information and place: One is not generally in the position to know that F (it is hot) without knowing where it is F (hot). The counter examples which immediately come to mind—that the information comes from the television, or that one wakes up on a desert island—serve rather to strengthen the case: for in such cases one knows quite well where the information comes from—"here" or "the TV." What one may fail to know is not where the object from which the information is derived is placed with respect to one's self, but rather where the object (the island, or the place depicted on TV) is placed with respect to some larger frame of reference (say, the globe). To use Evans' terminology: to gather information is always to place the origin of the information in "ego-centered space." Indeed, it seems unlikely that we would count any perception which we did not experience as originating (somewhere) in the world *as* information.

In this sense an object can be conceptualized not just as a collection of information, but also as a collection of places associated with that information. The sorting criteria for information must be intimately related to the sorting criteria for places, for the boundaries of the information-object are described by the boundaries of the placed, spatial-object. How, then, do we know—have epistemic access to—the spatial boundaries of objects? My eventual claim will be that our access to these boundaries comes not in the perceptual-information field,[62] but in the behavioral field of ego-centered space. But to make sense of the claim we need first to discuss the behavioral significance of space.

Let us begin with the example of auditory input. It is usually the case that we hear a sound as coming from some direction, or even from some particular position in space. The knowledge of the location comes to us with, but is neither reducible to nor derivable from the information regarding a descriptive characterization of the sound. Consider, in this regard, hearing a descriptively identical sound on land, and under water. Since sound travels faster under water, our directional sense is easily fooled, so that despite the descriptive (qualitative) identity of the two sounds, in the latter case we might not know where the sound was coming from.[63] (It is not uncommon, underwater, to hear sounds which seem to come from everywhere at once.[64]) Yet when we do place the sound, "the apparent direction of the sound is part of the informational state: part of the way things seem to the subject, to use our most general term for the deliverance of the informational system."[65]

Evans claims, quite rightly, I believe, that in hearing the direction of a sound, the specification of that direction comes (and cannot fail to come) in "ego-centric" terms, which Evans defines: "The subject conceives himself to be in the centre of a space (at its point of origin) with coordinates given by the concepts "up," "down," "left" and "right," and "in front" and "behind," and we may call thinking about spatial positions in this framework centering on the subject's body 'thinking ego-centrically about space.'"[66]

It is important to realize that the *content* of the subject's understanding of these coordinates and thus locations within them is not conceptual in the normal sense of the term—one needn't have mastery of these *concepts* (e.g., linguistic mastery of "up," "down," and so on) to indicate locations in space, nor to experience information in spatial terms. Evans notes:

> Some people, including, apparently, Freud, are able to understand the word "right" only via the rule linking it to the hand they write with. But when the terms are understood in this way, they are not suitable for specifying the content of the information embodied in directional perception. No one hears a sound as coming from the side of the hand he writes with, in the sense that in order to locate the sound he has to say to himself "I write with this hand" (waggling his right hand) "so the sound is coming from over there" (pointing with his right hand). Rather, having heard the sound directionally, a person can immediately say to himself "It's coming from over there" (pointing with what is in fact his right hand), and may then reflect as an afterthought "and that's the hand I write with."[67]

How then should we think about the content of this (ego-centric) knowledge where? Evans' answer is simple: "The spatial information embodied in auditory perception is specifiable only in a vocabulary whose terms derive their meaning partly from being linked with bodily actions."[68]

To anyone who has followed the work so far, the use I wish to make of this insight should be predictable: The content of a subject's knowledge regarding the direction and location of perceptual information comes not in terms of what we might call "objective space" (in which concepts like compass headings, distance measurements and landmarks might come into play)[69] but rather in terms of "egocentric space," which is, at root, a *behavioral space*.[70]

But this connection needs to be explicated more fully. To know where some object is ego-centrically is to be disposed to treat information gathered about that object as germane to one's behavior.[71] Evans notes: "One might, in the dark, be struck with the thought that there is something immediately in front of one's nose; if this was correct, then one might be disposed to respond in a certain way to information one would receive from it (the place) if the lights went on."[72]

This dispositional connection, as we would expect, is strong enough to suggest its necessity to place-directed thoughts in general. The content of a specification of location, when it is thought of ego-centrically, cannot fail to be behaviorally and dispositionally significant. Thus: "It is difficult to see how we could credit a subject with a thought about *here* if he did not appreciate the relevance of any perceptions he might have to the truth-value and consequences of the thought, and did not realize its implications for action (consider, for instance, a thought like 'There's a fire here')."[73]

This connection exhibits itself in the other direction as well: not only is placed information behaviorally significant (relevant to determining appropriate behavior), but behavior is relevant to placing perceptual information, for a subject's informational state is partly determined by the nature of the activity of the subject in the course of gathering that information.[74] Evans brings out this point in considering the case of tactile-kinaesthetic perception (in the absence of visual input):

> a blind person (or a person in the dark) gains information whose content is partly determined by the disposition which he had thereby exercised—for instance, the information that if he moves his hand forward and such-and-such a distance and to the right he will encounter the top part of a chair . . . we can think of him ending up in a complex informational state which embodies information concerning the ego-centric location of each of the parts of the chair.[75]

Here the subject, having gathered the information by reaching out in such-and-such a way, thinks of the location of the encountered object as the place he reached to, in terms of the behavior which generated the informational contact. The locations (being "there"), specified in ego-centric terms, thus gain their epistemic significance because of their relation to bodily activity, both the further dispositions to act with respect to the located object, and the activities themselves through which the information was gathered. To be "over there" is to be at a point in a behavioral space, accessible and specifiable comportmentally: reaching out (say) or pointing.[76] Our perceptual space is at root a comportmental space, centering on the body, and knowing a location in ego-centric terms is a behavioral knowledge expressed actively: moving "there," pointing "that way." Each location in ego-centered space has behavioral significance in this sense, and only thus is the spatial organization of perceptual information behaviorally relevant.[77] Our experience of space, and our experience of information as placed, is thus rooted in the body and its activity; our interpretive relation to information is mediated by our comportmental relation to the world.

4.23 Ego-centered and Objective Spaces
   I wish to stress the fact that this analysis does not deny the possibility or necessity of *objective* thinking about space. Indeed, I think this behaviorally grounded knowledge of space and its structuring of sensible information is at the root of our knowledge of objective, interpersonal space.
   It is clear, as Evans notes, that our recognition of ourselves as a bodily presence in space requires that we recognize that the space in which we are located does not depend for its existence on our presence. "We say that [a] subject thinks of himself as located in space (in an objective world that exists independently of him, and through which he moves); only if this is so can the subject's egocentric space be a *space* at all."[78]
   We cannot imagine a person having knowledge of the location of some particular object without relating that location to the self (this is precisely the condition of an *ego*-centered space) and thus possessing the capacity to think of the self as placed. But this is not placement of the self *in* ego-centric space in any simple way, but placement in a system of relations which is independent of one's particular perspective and understanding, i.e., it is a placement of the self in objective space. We have been inquiring here as to the basis of our knowledge/ understanding of space, but that such knowledge necessarily constitutes an individual *perspective* does not warrant or entail the conclusion that space is an artifact or construction of this perspective. Indeed, only if we think of our knowledge as an individual perspective (or, perhaps better, as an individual relation to an objective phenomenon) can we think of it as knowledge *of* space, knowledge of the relations in which objects necessarily stand to one another. We must be able to think about spatial relations not just in such a way that we can recognize that the lamp's relation to me is also an instance of the relation of the chair to the desk (and so think of the self as one object among others and not only as the origin of spatial coordinates) but also in such a way that it is not surprising, coincidental, or in need of explanation that two individuals approaching the same object (ego-centrically located for each of them) also approach each other, and end up at the same place (so that individual knowledge of space is related one-many to actual places, or that the "sense" a given individual attaches to a location in fixing on it is a perspective on the *same* referent-location to which another

individual attaches a different "sense"); we must be able to think this way about not just other people's locations, but also of our own. To fail to be able to make sense of space this way would, it seems, be to fail to make sense of space at all. To claim otherwise would be to claim to understand an exchange of the following sort: "Where did you see the rabbit?" "Oh, right over there, but not anywhere you could go." "No? Is it private property?" "Yes. It is *my* there, not yours, which I am referring to. You might go there in your sense, but then you wouldn't be where I was, and so you wouldn't see the rabbit, because he isn't there. He is there."

Something like the above might have made Lewis Carroll happy, but it hardly presents material for a workable conception of space, even space individually and indexically understood. Unless we can make sense of the two different personal uses of "there" having the same objective referent, then we cannot make sense of space at all, even indexically. If the person who saw the rabbit continued to insist on giving such "locations," we would have to conclude that the person didn't actually *see* a rabbit, or at least that the rabbit he saw (say in a dream) wasn't anywhere at all.

But there is also a more practical way in which ego-centric knowledge of space grounds objective knowledge: an ego-centric fix on the direction from which given information comes can allow one to locate the origin of that information (to locate the object to which the information pertains). Take, for instance, the case of a termite in a beam. As Evans notes, the information link one may have with this termite (hearing it eat the wood) can support a demonstrative identification of this bug, even where one does not know where it is objectively. For this information link, by placing the sound egocentrically, continuously for as long as the sound lasts, gives one the practical ability to locate the bug; one can follow the sound to a given location, a location which is theoretically (even if not always practically) accessible to others—as for instance if the upstairs neighbor is digging into the floor to the *same* spot to which you are digging through your ceiling.[79]

We see, then, how the comportmentally-based epistemic access to space which comprises ego-centric space is at the same time epistemic access to the interpersonal, objective space where material objects are to be found, individuated and recognized. With this established, we can move to examine the role this access to space plays in the sorting of objective information and the individuation of objects.

4.24 Comportment and Information Sorting
Two important connections have been made and explored so far. First, we have noted the important connection between our capacity to gather information, and the ability to place the origin of information so gathered in an ego-centered and objective space. Indeed, our inquiry seems to indicate that we can only make sense of having an informational relation to objects insofar as we understand the

objects and ourselves to be occupying the same space. Second, we have explored the close connection between ego-centered space and bodily activity. These two connections provide us with the resources to explain our capacity to individuate objects and to sort perceptual information appropriately, which explanation must be central to any account of the structure and significance of our sensory field, as well as to any claims regarding its veridicality and appropriateness.

Given the connection between information and place we might state, rather simply, that a sufficient condition for knowing of any information F to which object a it pertains, is to know the spatial origin of F and the spatial boundaries of a (or, if the object is too large to know its boundaries per se, to know the spatial emplacement of a is sufficient: where, in the epistemically accessible spatial range, a is). This capacity for sorting perceptual information according to the heres and theres of ego-centered space[80] is what underlies our capacity to individuate perceived objects ranging from trees to rainbows. Our ability to structure and arrange perceptual information thus depends upon possessing not just a sensible, but also an active, comportmental relation to the world.

But the importance of our embodied access to the world only begins with its role in our capacity to place perceptual information. Our access to physical solids also centrally involves bodily interaction: knowledge of material objects as such, of their materiality per se (our knowledge of their emplacement as material things) is rooted in the *behavioral* significance of our encounter with the object.[81] Restricting our analysis, for the moment, to the *solidity* of the object as the foremost sign of its being emplaced, we can say that the significance of its solid emplacement comes to us not as sensation,[82] but rather in its relation (usually resistance) to our activity as an embodied being.[83] Indeed, it is unclear that we even have such a category as "solid" for sensation; hard, soft, flexible, springy are all *ways* of being (a) solid, or ways of experiencing solidity. But I doubt that there is anything common to these experiences which classify them *sensibly* as solid-feelings. Rather, the solid is that which restricts our movement, that through which we may not pass. It would be misleading to say that the solid presents itself to us as resisting our will; we may without contradicting the solidity of the thing will the floor to hold us. Likewise we may will to walk through walls, and the intention will meet with no opposition. The locus of resistance for the solid is not our will, but our body; the solid is arrayed not against our desire, but against our activity.

If space is experienced around the comportmental potentials of the body, then our *actual* space is divided, arranged, and limited by the behavioral restrictions of the solid. The materiality of the thing is known by its limits to the activity of the subject, and of course these limits, encountered in a behavioral *space*, are also the spatial limits of the thing. Even at this level of crudity, then, we have accomplished a bit of our goal: for any information gathered, this division enables us to know from which object a the information comes. Of course, insofar as we have made a division only between the solid and the ethereal, we can say only whether the information originates from somewhere in the solid, or not. Our capacity to place information may allow us to individuate such visual objects as rainbows, but it is not sufficient to explain our capacity to individuate particular solids.[84] Thus it is to this capacity which we must now turn.

4.3: Individuals and Independence

What I have been claiming is that our access to things as individuals is provided by the active body, whereby our physical interaction with things reveals to us not just their boundaries as material objects, but also their individuality as independent things. We have noted already how the solidity of the thing can serve as the primary sign of its being emplaced as a thing (and to what can we attribute the boundaries of the material object, displayed as resistance to our bodies, but to the *thing*, here asserting itself as existing over-against us, independent of us?), such that to fail to be available as solid is thereby to fail to *be* a material object. But we must ask how it is that the thing can display its boundaries as a marker of its independence from *other* things.

My suggestion is that an individual[85] is such in virtue of its availability to practical alteration independently of its surroundings, and I mean by this not that the manipulation of the object in question would leave its surroundings unaffected, but rather that the *particular* character or nature of the alteration would be limited to the individual; the different ways in which the alteration would reveal itself in the object's surroundings would itself serve to mark the independence of the object.[86] Spatial manipulation is a central example of the sort of practical alteration which I have in mind here. To be an individual is to be independent in this sense: an individual can be moved without otherwise altering the configuration of behavior-restricting solids;[87] it is available to manipulation in isolation from surrounding objects. This is a particular example of what I take to be a far more general phenomenon: our recognition of material objects as such involves or requires our bodily insertion into the causal order, an alteration of the world which may reveal to us that which sensation will not.[88] In this case manipulation of a thing reveals its boundaries not just because such interaction involves encountering as solid the object in question, but because the manipulation alters the spatial-behavioral configuration of our environment in a very specific and limited way without having similar consequences for (formerly) adjoining surfaces. The extent of the alteration reveals for us the extent of the thing. To grasp the edge of my pen and move it is to discover the boundaries of the pen both as the solid which resists my grasp, and as that solid volume which, in being removed and replaced to that extent, alters the experienced spatial configuration of my environment. Of course this movement is also observed (in recognizing the spatial boundaries of material particulars one can thereby sort sensory information) but it must be insisted that access to boundaries in this way does not have the epistemic status of mere observation;[89] the consequences of the manipulation of material particulars are always also consequences for further action and manipulation. As the spatial configuration of the material world changes, so change our opportunities for activity. Thus our access to, and the sign of, a thing's unity as a material particular, its spatial boundaries as an individual, comes in interactive behavioral space, even as those boundaries are represented in the sensory field.

Here we can begin to see the ways in which the relation between representation and reality (between "object" and "thing") is cemented by the living, active body.[90] The principles of division of the sensory field are grounded in our comportmental encounter with the world; by knowing (actively) the spatial boundaries of the individual, and the spatial origin of perceptual information, one thereby possesses the resources to divide appropriately the sensory field. This capacity to sort perceptual information by object signals an epistemic sensitivity to the physical object which is not merely a sensitivity to its sensible attributes— for on this conception the object is not merely a collection of attributes, but an emplaced solid which is not as such reducible to, nor available in terms of those sensible qualities.

4.31 Manipulation and Degrees of Freedom

So far the examples we have been dealing with have been extremely simple: as in a small, solid cube sitting upon a flat surface. We have seen how we might know to sort the information coming from the two solids (cube and surface in our simple picture) from other information, given our twin abilities to know spatially the origin of information and the boundaries of solids.

And we have seen the role played by physical interference in such a two-body system in discovering independence; we grasp the cube and pick it up, thus displaying both visually and behaviorally its independence from the surface.[91] Such interference is effective precisely because movement or manipulation reveals the boundaries and extent of the thing in terms of the limited alteration of the behavioral field which results.

But the two-solid example is simple in a number of ways. First, the surface is considered to be stable enough, and the surrounding environment empty enough, that there is no other reaction to the manipulation to interfere with one's interpretation of the effect of that manipulation. Second, the cube is a rigid solid, so the reaction of the whole to the interference was uniform, it all moves in just the same way and direction as the grasped sections of the solid, without physical distortions which would effect the way in which the solid volume would be behaviorally experienced in its new location (or along the way).[92] And third, the cube is a simple individual, in that it is possessed of all six possible degrees of freedom: it can move laterally as independent object along all three spatial axes, and it is free to spin around all three rotational axes. This means for the manipulator that the actual manipulation can take place along any of these axes of movement; the only freedom of the manipulator restricted in this example is that restricted by the solid itself: i.e., it is incompressible.

I do not plan to spend time discussing complications of the first and second sorts, for establishing the particular and no doubt fascinating complexity of our recognition of individuals, which perhaps often involves a complicated interplay between behavioral and perceptual clues, is not the purpose of this work. I wish only to show the importance (indispensability) of our interactive relation to the world to any explanation of individuation, and thus to establish the active body as a means of knowing. Nevertheless, the third complication deserves some notice, as it seems to be the root of our recognition of different kinds or degrees of what is nevertheless individuality.

Let us again consider some examples to make the contrasts apparent. The cube from above is an easy example of a simple individual: its particularity is defined by its boundaries, and their potential separateness from (their potential to be noncontiguous with) all other solids. Not so for the handles on the bathroom faucet. Such individuals are bound by structure to be always attached to surrounding solids (say, the countertop). And yet it is nevertheless practically important to be able to recognize their functional individuality. Here that individuality is recognized not via a display of its independence along axes of lateral motion (although of course handles can be removed as a sign of their independence as objects, it is only at the expense of their functional attachment to the mechanism as *handles*) but rather through realizing that the handle is amenable to rotation. This freedom marks it as independent in an important way (although in an equally important way quite integrated into the larger whole of which it is part)[93] and thus, in that way, as an individual material particular.

It would not be wrong to understand these criteria of identity precisely as invitations from the world to divide, remove, turn, breathe, or otherwise manipulate the objects of the world according to, as is appropriate for, the nature of the objects in question. The cube, being a unit, does not invite spatial division in the way which the leaf, being itself naturally divided, invites separation and recognition of those parts as such. *It is through inviting and directing the activity of the body that the world guides our recognition of its inherent structure.*[94]

4.32 Physical Boundaries and Visual Cues

It must absolutely be acknowledged that we often, perhaps even usually in our adult lives, identify boundaries and edges, and by extension individual particulars, with visual clues thus:

figure 4.1

With just simple arrangements of angles, or shadings, we identify the outlines of solids and determine the individuality of objects. I would be foolish to dispute this fact. My contention, however, is that the significance of these visual clues is not inherent in their visual qualities, but comes instead from the fact that these "looks" have come to be signs for things known independently, behaviorally in the case we have been arguing. No doubt these perceptions have behavioral significance, and they surely have (let us call it) metaphysical significance in so far as they allow us to identify particulars, but this significance cannot be derived from the qualities of the perceptual field considered as such; there is nothing intrinsic to the phenomenological quality of the shading around a sphere which makes it signify the boundary of that sphere.[95]

We might borrow some implications from Wittgenstein's "private language argument" to help make the point. I take one upshot of his remarks to be that the notion of sensations or perceptions with intrinsic qualities defining their content, which content is the basis of linguistic meaning or epistemic significance, is a bogus one. Such qualities, if they exist, play no role in the content of sensation or the foundations of meaning. However, this does not imply that there is no content to sensation at all, nor that this content is unrelated to knowledge (and it certainly does not imply that there is no epistemically significant perception). We have simply misunderstood how it is that our perceptual experience comes to have epistemic significance for us.

Wittgenstein clearly insists on the importance of sensations. He suggests, even, that we might explain to someone the meaning of the word "pain" by pricking him with a pin—causing a sensation and naming it for him—"see, that's what pain is!"[96] But his idea is most certainly *not* a return to a kind of sensational nominalism, for Wittgenstein intends us to understand this relation between perceptual experience and epistemic significance as being quite the reverse of the relation suggested by that doctrine. As is clear in his discussion of the diary for sensation "S," such a sensation, whatever it is, comes to have content—the experience comes to have significance—only insofar as it comes to play a role in our lives and activities. In particular, such a sensation is significant exactly insofar as it comes to imply, be related to, or becomes involved with, my understanding of a particular event, behavior or practice. When I note that "S" occurs when my heart monitor indicates an increase in pressure, it comes to be the way my rising blood pressure feels. The significance of sensation is captured entirely by references to such empirical concepts as "rising blood pressure," or "pain"—and, as a result, the content of the experience is likewise determined by the concept of the thing experienced. Wittgenstein writes:

> Let us now imagine a use for the entry of the sign "S" in my diary. I discover that whenever I have a particular sensation a manometer shews that my blood pressure rises. So I shall be able to say that my blood-pressure is rising without using any apparatus. This is a useful result. And now it seems quite indifferent whether I have recognized the sensation right or not. Let us suppose I regularly get it wrong, it does not matter in the least. And that alone shows that the hypothesis that I make a mistake is mere show. (We as it were turned a knob which looked as if it could be used to turn on some part of the machine; but it was a mere ornament, not connected with the mechanism at all.)
>
> And what is our reason for calling "S" the name of a sensation here? Perhaps the kind of way this sign is employed in this language game. —And why a "particular sensation," that is, the same one every time? Well, aren't we supposing that we write "S" every time?[97]

Now, statements such as "It feels like my blood pressure is rising," "it feels like a pinprick" or, especially apropos here, "it looks like an edge" are perfectly natural (and it is important that this is so), but because of this I think that it is easy to miss the philosophical significance of understanding the meaning of such statements in the way outlined above. There is no sense we can give to the "like" in the above statement which we could understand independently of our conception of rising blood pressure or edges; there is no indication of the existence of experienced qualities of the sensation which mediate the identification of the sensation with rising blood pressure (so that, for instance, the quality of the sensation is the immediate or primary content of experience which one has learned to identify with some object or event). What is primary in perception is not sense data or qualia or some other characteristic to the sensation/perception (its intrinsic "feel"), which the perception has independently of, or considered apart from, the concept of the thing so perceived; what is primary to perception (what makes it the [kind of] perception that it is) is precisely the concept of the object perceived, where this is understood in terms of the criteria of identity for that object.[98] ("And what is our reason for calling 'S' the name of a sensation here? Perhaps the kind of way this sign is employed in this language game. —And why a 'particular sensation,' that is, the same one every time? Well, aren't we supposing that we write 'S' every time?"[99]) Even the criteria of identity (sameness over time) for the sensation is to be understood in terms of the applicable criteria for its associated concept—don't we write the same letter each time? Wittgenstein's critique of empiricism reveals the contents and significance of our sensations as deriving from the significance of our concepts. What makes a given thing "look" like an edge, or a boundary, or a separable thing, is nothing about the "look" considered in itself, but is precisely the fact that it is a perception *of* an edge. It is the way an edge—which, in marking the bounds of a solid has some very particular behavioral significance, and in allowing us to sort information by solid has epistemic significance—looks. What it is to *be* an edge is defined and known in terms of its role in our active engagement with the world; what it is to *look* like an edge is parasitic on this prior access to its identity as such. Thus, the epistemic significance of the perceptual experience of a boundary derives from the role which a boundary plays in our lives; that which grounds the identities of boundaries and that which determines the "look" of a boundary together form our idea of the boundary.

4.4: The Body's Access to the World
     The effect of the foregoing discussion cannot be to deny that perception, in the traditional sense whereby the impacts of the world on the sense organs are causally transmitted to the brain/mind for interpretation, is a central mode of epistemic access to the world. What must be denied at this point is only that the transmission of these impacts can dictate the content of the perceptual judgements which result. For causal-perceptual signals have significance only insofar as they are taken up into an already formed, and relatively stable conceptual structure. As we saw briefly in section 4.32, we needn't deny that perceptual data of this sort have significance; rather we need to see how they came to have the significance they do.[100] Part of what I have been arguing so far is that an appearance or description can act as a criterion of identification for a particular only in light of its relation to behaviorally accessible criteria of identity,[101] known through the relation and response of a thing to practical physical involvements. In this way our comportmental access to a thing influences the formation of the concept(s) whereby perceptual information is interpreted; the shape and structure of the physical world limits and directs bodily activity, and thereby influences the concepts by which we interpret our percepts. Thus our use of the appearance of a thing as a criterion of identification depends upon our behavioral access to the identifying boundaries of that thing.
     It might be objected that the account I have given here amounts only to the claim that our sense of touch is of central importance to the whole of our perceptual contact with the world; and while touch is often devalued in favor of more visual metaphors for knowing (and so perhaps correctives are always in order) still the centrality of touch is a long recognized fact of perception. But such criticism would not be entirely appropriate, for insofar as our sense of touch (our skin qua sense organ) is itself a causal mechanism, then the impulses it sends to be interpreted into conceptually significant content are in the same epistemic boat

as the deliverances of the other senses. In touch-perception there is just the same severance of cause from content, and therefore just the same vulnerability to the skeptical wedge driven between understanding and the world. The world is equally unable causally to guide or limit the content of tactile experience in a way which would justify the claim that we are thereby "in touch with" the world.

Thus just so long as it is made from within the broad empiricist framework which it is my purpose to challenge, the claim that tactile-sensation possesses a central organizing role in our perceptual awareness as a whole, however justified in itself, fails to address the central problem of epistemic-*openness*, of the capacity which allows the world to guide and limit our conceptions of it.

What I am in fact claiming is that our tactile-*kinaesthetic*[102] sense has a central organizing role for perception as a whole, *and* that the particular epistemic sensitivity afforded us by bodily motion allows the world to limit and guide our organizations of sensation. It is a central advantage of this account that it allows us to maintain that our identifications of objects needn't (by definition) coincide with their actual identities—thus acknowledging fallibility—and yet insists on a form of epistemic openness to the identities of objects—thus avoiding an immediate fall into skepticism.

We have seen already that the claim to be in touch with, to have access to, some particular depends on the possession of an openness to the world whereby that particular can provide some epistemic friction by which our experience can be guided. What I am arguing is that our empirical concepts (which determine the content of perceptual experience) are open to the world, via bodily activity, in just this way. Through activity the world can guide us; because bodily motion must be conceived as taking place in the world, activity is not beyond the reach of the world, and the spatial and material structure of the world has definite implications for motion. The concepts most at play in structuring our perceptually-rooted account of the world, such as place, size, weight, solid, edge, individual, and sort, are significant because and in terms of the possible, available and appropriate motions of the body. Should our conception of the world indicate a wall here, an open vista before us, and a small object on a desk, such judgements have implications for our range of motion, and it is through its limitations on our motion that the world can correct any misconceptions: so we might try to walk into the vista before us only to discover painfully the existence of an especially clean glass door. Certainly it would make sense to attribute to the world the change in our understanding brought about by smacking full into a sliding glass door, and it would be possible to make this attribution, as it was not under Davidson's theory, in virtue of the epistemic access to the world provided by the kinaesthetic-body.

Because the active body is necessarily always open to the world in this way, our conception of the world is always open to revision.[103] Not only can the influence of the world on our bodily activity often account for the fact that one concept rather than another was drawn into operation in our perceptual experience, but because the active body is necessarily always open to the world in the way detailed above, the concepts by which we interpret the world are always open to revision. This means in part that change in our conception of the world before us from "open vista" to "glass door" is to be attributed to the world's influence on our behavior in the immediate sense, and in part that the divisions of our conceptual matrix in terms of which our understanding of reality is cast can themselves be elucidated in terms of bodily activity. As it was part of this section (and will be part of the next section) to show, the meaning of such terms as "solid," "mountain," or "water" can be given in comportmental terms: motion ceases at this point, or is restricted there; here a different technique of bodily motility must be employed. Concepts are in part constituted and defined by their comportmental significance, and it is largely in virtue of this comportmental significance that they can function empirically in perceptual experience. But because our concepts are comportmentally significant, the further investigation of particulars as such, or as members of given kinds, can lead to the discovery of thus far unknown comportmental significance. Thus the body's openness to the world implies the revisability not just of our conception of the world as a whole, but also of the terms in which that conception is cast.

It is important to realize that I am not proposing that the body and its activity are (perhaps especially reliable) causal or epistemic intermediaries between concepts and the world. Such a proposal would structurally mimic that of the traditional or pragmatic empiricist by inventing a kind of experience to which our understanding could converge. But we have seen that this only raises the unanswerable question of the reliability of this experience, its claim to represent the world. As Davidson and McDowell have made clear, it is important to recognize that there are no such epistemic mediators between our concepts and the world. Activity does not possess the indeterminate relation to perceptual content characterized by cause, nor does it occupy the epistemic position of contentful experience impinging on our comprehension. What I am claiming instead is that the mode of our conceptual openness to the world is bodily; it is because of the body's place and activity in the world that our empirical concepts are themselves receptive to, and open to revision in the face of, the material structure of reality.[104]

One upshot of this ability to claim a nonperceptual openness to the material structure of reality is the rehabilitation of realism. The epistemic resources of the kinaesthetic-body reveal the permissible orderings of the sensory field to be more restricted and better grounded than the relativist or idealist would have us believe. Determinations regarding the appropriateness of our conceptual schema to the world are thus possible via the cement of the body, which ties our representations to the world through the conduit of agency and activity.

Section 4.6 will take the question of the role played by our embodied activity in the world in grounding our perceptual criteria of identification for sortal types.

I will argue that we can, and do have access to natural kinds because our epistemic sorting is the product of active interference *in* the world, and not merely the product of active thinking *about* the world.

4.5: Of Transcendental Arguments and Swamp Men

I claimed as early as chapter 1 that I intended to advance a transcendental thesis: being embodied, and having comportmental commerce with the physical world, is a condition of intentionality. Before moving on to the argument promised above, I would like to address briefly two strains of objection to this transcendental claim, and thereby clarify and explicate the different empirical and transcendental implications of my argument. The first sort of objection, well known from the work of such figures as Barry Stroud, purports to show that no transcendental arguments about the nature of experience can serve as an effective critique of skepticism. The second sort of argument I will address here comes from Donald Davidson's Swamp Man: this thought experiment purports to show that the causal history of our interaction with objects does not matter to our knowledge of them nor to our capacity to refer to them, since Davidson can imagine an exact molecular duplicate of himself coalescing from a freak arrangement of swamp gas, and we presumably would not deny to Donald "Swamp Man" Davidson the capacity to know and refer to the same objects which the original Davidson can. (They are, after all, molecular duplicates.)

The general thrust of the Stroud argument is that transcendental arguments about the nature of experience can at most show that we must take the feature of experience in question to be just that: a feature of our *experience*. Thus, if I were to construct a successful transcendental argument to show that we cannot make sense of our experience except as the experience of a real, physically solid, spatially extended world, Stroud would counter that all I have shown is that this interpretation of the import of our experience is a necessary feature of that experience. But like all experience, it may turn out to be false. Stroud exploits the experience-reality distinction (which must, of course, be preserved in some form) to undermine transcendental argumentation regarding the necessary structure and contents of our experience; the very element of doubt about the reflection of reality by/in the content of our experience which gives rise to skeptical worries in the first place—the realization that the world does not determine content—can be used over and over against *every* claim about experience.

Now, without recourse to the generally accepted and apparently banal thesis about the possible slippage between experience and reality—we might call this the thesis of the autonomy of content—the Stroud argument would look unmotivated. The thesis of the autonomy of content has at least two posits: (1) experience may fail to "match" (or be appropriate to) reality, and of course this is unobjectionable (for it must be true of anything which counts as experience that it could fail to reflect the world). And (2) that which *determines*—shapes, alters, limits and directs—the *content* of experience, may be not just *accidentally* (as with insanity) out of touch with reality, but *naturally*, as a function of the structure of the experiential system, out of touch with reality. That is, the skeptic must hold that our epistemic system is *structured* in such a way that as a result of its normal functioning, content could become radically and *unrecognizably* discordant with reality.[105] I am not claiming that the skeptic assumes what he is meant to prove: the skeptic may believe that we are in actual physical, causal contact with the world, and may even think that our beliefs are true (even if verification transcendent). What he will deny is rational warrant for that belief, and he will have to point in the course of his defense of this stance to some version of the thesis of the autonomy of content which has both provisions.

What *my* transcendental argument about the nature and structure of experience questions is precisely the second posit of the thesis of the autonomy of content. I don't deny that sensory inputs require conceptualization (that is, I am not denying the autonomy of content from the bare causal determination of sensory inputs), but I am claiming that the very concepts at play in synthesizing sensory inputs into the stream of experience are tied to the world via comportment, and tied in a way which, if not fixed, is severely restricted. The precise nature of the connection, and the functional relations between comportments and the articulations of a representational space, is an empirical matter, which, not for lack of interest, I am not addressing here. I am interested here in establishing that on such a model of experience we have some justification for, and some guide to a theory of, the veridical connection of mental states to the world: we could be justified in believing that, and have some understanding of how and why, my thought about him is, indeed, about *him*.[106] And, further, the model I am offering allows for the possibility of mistakes while at the same time showing that we have an open epistemic conduit to the world which can serve to correct our empirical misapprehensions. It may of course turn out that skeptical doubt can be raised on this model too (not the normal local uncertainties which are a necessary part of our being in the world, but global doubt, which involves denying in some sense that we are fully living in the world[107]), but it will take different sorts of arguments than are usually leveled at the realist. To show that something is a feature of experience is only sufficient to place it back under the umbrella of doubt if we assume the full autonomy of content, and it is just this which I am denying.[108]

Now the swamp man argument is of a different sort, and is in fact constructed to counter arguments which, while not unrelated to mine, are also not the same as mine. Thus the objection will take some reconstructing. But the basic idea is that, since we can imagine the possibility of subjects with entirely different causal histories nevertheless possessing the same intentional ties, we cannot make causal history a transcendental restriction on intention. This seems a perfectly benign thesis to me, since I do *not* claim that in order to refer to or think about a given particular one must have had actual physical interactions with that object. What

I *do* claim is that intentional connection is impossible outside of epistemic openness, outside of the epistemic availability of the particular in question. And epistemic openness requires our active, bodily sensitivity and responsivity to the world and its objects. Philosophers with a different mind-set and more patience might be able to spell these conditions out in terms of subjunctive counterfactuals: if the intended object *were* present, then the epistemic resources of the subject would be sufficient to track the successive states and positions of the object and suitably alter the connections between desires and the behaviors intended to satisfy them. If the cup of coffee I am thinking about were present, and were moved or spilled, the same epistemic openness which allows me to *think of* that cup (is a condition of my intention) would also provoke a shift in my behavioral attitudes consonant with the changes in the state of the coffee; I would reach *here* instead of *there* to get it, and I would not reach for it at all (to get coffee) if it were on its side and emptied. These behavioral attitudes would, in part, constitute my knowledge of the positions and states of the coffee.

I claim that it is an *empirical* fact that much of our knowledge is gained in the course of our physical negotiations of the world, and that much of our knowledge is, indeed, constituted by these interactions, and lays the groundwork for absorbing and interpreting knowledge gathered by other means (by hearsay, broadly construed). And certainly much of our knowledge *is* gathered by other means: the vast majority of our formal education is education by hearsay, after all. It seems to me unwise to make the transcendental claim that causal history determines or restricts the range of our intentional connections (even if we elaborate this basic thesis with the idea that we utilize the causal histories of others in learning by hearsay, thus somewhat eliding the difference between these modes of gathering knowledge) since there is *prima facie* convincing empirical evidence to the contrary.

The transcendental claim that *I* am making is different and more restricted: epistemic openness to (the epistemic availability of) particulars (and *not* our actual involvement with or the current presence of particulars) is a condition of intentional connection to them.[109] And, of course, I claim that the capacity for bodily interaction is a condition of (indeed, is in a sense the locus of) epistemic openness.

This raises a host of issues about the epistemic limits of the physically handicapped, which it may be worth addressing in some detail (this could take us into a fruitful discussion of Merleau-Ponty, whose work exhibits an unrivaled phenomenological sensitivity to the myriad ways in which our bodily *being-in-the-world* underlies our conception of that world. That bodily *activity* underlies knowledge gathering of every sort—from the scanning movements of the eyes to the adjustment of the position of the head in hearing to the grosser movements of the limbs in everyday physical navigation—is only the simplest form of this claim). But I shall not undertake such a project here. As far as I am concerned, it is obvious that the vast majority of even the severely disabled possess the requisite bodily-based epistemic openness which underlies intentionality. In the case of cerebral palsy, for instance, the range, fluidity and predictability of muscle movement is severely restricted, so that in the course of reaching for that coffee cup, a muscle spasm might cause one to reach *through*, instead of to, the cup, knocking it over. But such a mistake is hardly grounds for the claim that the person was mistaken about *where* the cup was, or what actions were required to reach it. The motions of the disabled are no less a negotiation of the world than are the more predictably smooth bodily motions of the fully-abled, and they provide to the subject information about the structure of his or her behavioral space in the same way, for there is no reason to suppose in this case that the behavioral system lacks the appropriate links to the informational, sensual and intentional systems. What might be frustratingly challenged is the ability to conform those movements fully to the experienced shape of one's behavioral space.

In the particular example of cerebral palsy it is worth keeping in mind that only some motions are uncontrolled, and often only some of the motion involved in a given action is uncontrolled. But what of the epistemic status of the uncontrolled action? I suspect that the epistemic value of the truly uncontrolled action is nil, and that we should say that the person whose bodily motility was entirely uncontrolled (or who had no muscle movement, and whose motility was thus under the control of another) would *not* possess an epistemic openness to the world sufficient to ground intentional connection. But we should be clear just how radical a case this would be: first of all, even the simplest motion requires the coordination of not just the muscle fibers which make up the muscles involved in the movement but of several muscle groups, which must either shorten (the biceps in the case of the folding of the arm) or lengthen (the triceps) in unison to make the motion possible. A person whose muscle firings were truly random would not be capable of any recognized major motion or action. But we can consider the less radical (but in its way far stranger) case of someone capable of recognizable major motions and coordinated muscle contractions, but for whom these motions are entirely uncontrolled and apparently random. Such an individual would be capable of nothing deserving the term "exploration" or "negotiation" of the world, for his contact with the world would consist of partially coordinated flailings which even if they provided sensory stimulation would not be systematic enough to support anything like a settled sense of the physical structure of the world and one's place in it; there can be no behavioral space without *behavior*. Further, the uncontrolled actions would extend to the taken-for-granted movements of the eyes, which would thus be incapable of sustained focus or attention. It is impossible to imagine a life of what could only be considered random sensory stimulations—blurred colors which whiz by the eyes, and tactile stimulations which occasionally occur in the extremities—and grant to this being a sufficiently sophisticated epistemic openness to the world to support epistemic attention or intentional connection to the world's particulars. If you, dear reader, even with your life of epistemic connection and settled conceptual structures, were connected to a diabolical roller coaster, in which you lost control not just of your position and speed, but of the movement of your limbs and eyes, how long before you lost sufficient epistemic connection to the world through which you moved to lose, too, your intentional connection? In fact I think that in such a scenario (as in the case of an accident or illness which caused the loss of *all* voluntary motor control) one's epistemic access to the world would be so severely diminished as to involve the immediate loss of intentional connection. I must emphasize again just how radical a situation we are dealing with: To lose control of *all* voluntary motion means the immediate loss of all communication (no speech, writing, or expression) and a great deal of one's sight (no capacity to focus or to scan) and a corresponding loss of "scanning" in the sense of touch (textures would be difficult to experience). Taste might well be affected, since it would be impossible to eat (and to move food around the tongue) and smell, too, since it might not be possible to direct air through the nose. No doubt some senses in some situations would provide precise enough stimulation to call into operation empirical concepts allowing for correct interpretation of those stimuli, but (for reasons which have occupied my attention for many pages, and which bear no repetition here) in the absence of comportment, these concepts would not be open to the world. And, as I have been at pains to argue, without this openness we cannot claim to be epistemically or intentionally in touch with the world. Only if you believe that reference and intention is guided by, or rooted in, the accuracy or appropriateness of the mental images of the world which you might retain (as a result of your possession of empirical concepts which are by luck still appropriate to the world) can you insist that your intentional connection with the world remains once your comportmental-epistemic connection is severed. And what if you *never* had sufficient control over your movement through this world to establish epistemic connections, and the settled conceptual structures which make comprehensible sensual access to the world possible?

But it is only in the very radical cases of the person without control of muscle movement, or with no muscle movement at all that we can say that intentional connection is missing. Each case outside of the radical must be considered according to its particulars: to consider briefly one example of how one might proceed in such cases, it is not clear that we should attribute to the blind man an intentional connection with the rainbow which he has been told is glimmering over the mountain range ahead. He may well say to his young niece: "Isn't that a beautiful rainbow?" and she will know how to interpret his sentence, but without an *epistemic* connection to that rainbow his sentential reference may not signal an *intentional* connection. The problem is not that he cannot possibly have a notion of that which he has not experienced—language does for him in this case what it does for everyone in others, provides him with a concept which has its place in his mental repertoire—but it is not possible for him to have an epistemic connection with rainbows which is unmediated by another perceiver. If it is true that referential capacity requires epistemic connection, then a person with this restriction will only make genuine references to rainbows if he can use another person's direct epistemic sensitivity as a tool. It is not that he genuinely refers to rainbows when they are there, and fails to refer when they are not there, but that he cannot refer to them *even when they are there* unless he has arranged for himself an alternate epistemic conduit. There is a great deal of room for interesting discussion of such cases, not least of the conditions under which such alternate conduits are possible and sufficiently reliable for the purpose we are considering. Note, too, that the example is not directly applicable to the case of comportmental disabilities; sensation is a mode of epistemic receptivity, but, not, on my account, a mode of epistemic *openness*. It is clear that we need both sensation and comportment for knowledge, but (if I am on the right track with my analysis) each fulfils a different epistemic function, and thus malfunctions in each mode will affect us differently.

One important implication of the arguments made so far, an implication which will be more fully spelled out in the last sections of this work, is that brain-in-a-vat thought experiments are inappropriate in epistemology. If I am right (or at least heading in the right direction) then no brain-in-a-vat, no matter how carefully engineered its electro-causal connections, possesses the right sort of epistemic access to the world to qualify for epistemic openness or intentional connection. It is no surprise that skeptical conclusions follow from such a thought experiment: if *I* were a brain-in-a-vat, I would certainly be a skeptic! But I am not, or rather, I should say, the image of a brain-in-a-vat does not present a compelling metaphor for my epistemic situation. Thus, it is not clear that epistemic conclusions which follow from the situation of these poor beings should have any direct bearing upon me.

4.6: Interaction, Knowledge, and Natural Kinds

In chapter 1, I spent a few pages explicating Aristotle's view of knowledge and reality, and indicated my essential agreement with the Aristotelian line. It should be stressed at the outset, however, that my view is Aristotel*ian*, not Aristotle's; I am committed to the idea of a natural functional ecology which eschews any commitment to a strict substantial teleology. This is to say that the functional niche which, I will claim, defines each thing in virtue of its sort, and which therefore can often be explicated by teleological-esque "what-for" statements (chlorophyll is for photosynthesis), is a product of causal-functional relations internal to the physical-ecological system, and not, as in Aristotle's metaphysical cosmology, a purpose attributable to the individual in virtue of an external relation to a Prime-Mover or Being-qua-Being. My Aristotelianism is in this sense thoroughly secular and incorporates the structural metaphysics of modern science.

The notion of natural kinds which I will defend is not, however, the usual scientific-taxonomic one, rooted primarily in classification by occurrent property. This sort of taxonomy is a hold-over symptom of a sensual-descriptive epistemology which I argue we should discard, for it naturally leads to an epistemically closed internalism according to which we have no epistemic access to, and therefore cannot refer to, natural kinds. I have already discussed the adjustments we need to make in our epistemological assumptions to account for our individuation of material particulars, as the doctrine of sensualism turned out not to have the resources to ground the individuation of particulars necessary to comply with Russell's Principle. And we have seen, too, that it cannot ground a convergent realism. But I will argue that we can expect to converge (or in some cases can assume we have converged) on a real ontology if we accept the epistemic importance of our fully embodied, active presence in the world.

4.61 Heidegger's Hammer: Rethinking Epistemology

It is only fair at this juncture also to attribute the account of kinds and our access to them which I will give to Heidegger, or, more precisely, to my own encounter with Heidegger.[110] Heidegger is well known for his—I take it to be a very Aristotelian—critique of Cartesian metaphysics and epistemology, according to which "we are seen as fundamentally observers collecting data about the world through the senses and forming beliefs about objects on that basis."[111] In its place, Heidegger suggests a theoretical recovery of the more basic human experience of being in, and coping with the world. "The kind of dealing which is closest to us is as we have shown, not bare perceptual cognition, but rather that kind of concern which manipulates things and puts them to use; and this has its own kind of 'knowledge.'"[112] For Heidegger, as for Aristotle, our agency is fundamental to our status as thinkers and knowers. Embodiment is not something theoretically inessential to mindedness, as if the process of cognitive development proceeded without regard to, or in spite of, our existence as active, physical beings.[113] Instead, to paraphrase Maxine Sheets-Johnstone, thinking is rooted in doing.[114] Observing and thinking about the structure of the world does not precede interfering with that world.

> From Heidegger's standpoint, the concept of the knowing subject, trapped within its "veil of ideas" and constituting its world out of meaningless "hyletic data," is a highly specialized and refined way of understanding man which originates solely from epistemological interests and has no real counterpart in our actual lives. . . . In response to this traditional picture, Heidegger tries to show us that "knowing" [in this sense] is a "founded mode" of Being-in-the-world . . . [by focusing on] ordinary agents involved in mundane practical situations.[115]

What I am arguing is that one way to follow out this insight is by treating physical interventions in the world—everyday activity—as epistemically significant, involved with and not independent of, but not reducible to nor simply dependent upon sensation and cognition. One implication of this position has been explored already: insofar as thinking about particulars depends upon individuation, and individuation depends upon intervention, intervention is in this sense a condition of thinking.

But the force of Heidegger's insight also extends to a revaluation of knowing itself. Knowing need not mean theoretically connecting and conceptually synthesizing sensually gathered data to form a picture of reality. Such a picture analyzes our relation to the world in terms of our sense-organs' epistemically neutral *causal* connections to reality; but this ignores the epistemically significant contact of active intervention in the world. To accept the latter is to suggest that knowledge need not be a cognitive mapping of the known, for it can also take the form of the negotiation of the known. This is a trivial point to anyone who has ever (for instance) driven in Boston: there is a great gulf between the capacity to negotiate the streets of the city, and the ability to map them. To know Boston, where things are, is for most of us limited to the former. Indeed, I rather suspect that what cognitive maps we do have are for the most part inaccurate.[116] And yet we still manage (usually) to negotiate and navigate.[117] But Heidegger is making a claim stronger than this near triviality: "Understanding that something is such-and-such, or believing that some proposition is true, is impossible without understanding how to perform various actions or use various entities."[118]

It is not just that practical knowledge might be possible without explicit theoretical competence, but that theoretical knowledge would not be possible for a being without practical competence. No one who couldn't, at least in theory, negotiate the streets of Boston could interpret, much less produce, a map of those streets. Empirical-cognitive maps are rooted in practical negotiations. Here again, the current investigation has already led to similar conclusions: our understanding of objective spatial relations is dependent on our behavioral space. Likewise, the concepts which help structure our empirical "map" of the world are in various ways open to, bound up with, and dependent upon our embodied relation to the world.

Heidegger crystallized these various insights with his famous workshop metaphor. Our epistemic encounter with the world of the workshop is not accurately portrayed in terms of a subject confronting an array of *objects*: observing, inferring function from observation, and interfering in the service of observation to produce further observable phenomena to confirm or adjust the original "theories" in terms of which I made sense of the workshop environment. A more appropriate model is one of mindful engagement with the workshop environment, an encounter with the things of the world in which intervention, manipulation, and use constitute a knowing relationship with these things. "For this reason, Heidegger calls the mode of 'sight' in everydayness 'knowing one's way around' in contrast to the 'mere seeing' of the contemplative attitude. This know-how is a generally tacit 'feel' for the equipment at hand rather than an explicit knowing-that."[119]

We encounter the hammer, to use a favorite and oft-recounted example, in the context of hammering, of manipulating it in the service of certain goals. Our skill and ease with the hammer will be a measure not of our acquaintance with the thing as sensual object, but of our familiarity with the full range of its functional capacities and causal tendencies. The builder who drives a nail with one blow, then easily reverses his grip to slide the handle into its place in his belt, *knows* the heft and aim and balance of the hammer, can measure with the precision of his own movements the spin of the falling hammer, released from one grip to be caught in another more appropriate to the new aim of holstering. What the expert knows is what is acquired in the course of our everyday, often goal driven, sometimes exploratory, active engagement with the world.

Indeed, Heidegger's choice of hammering as a central metaphor for our interactive engagement with the world is in certain contexts especially appropriate. Anyone who has spent any time with infants and toddlers knows how basic an exploratory strategy hammering is; as soon as a child can grasp a thing, it has two simple tendencies: to put it in its mouth and to use it to hit things. The child's discerning exploration of its world from the very start involves not just sensual and shape-discovering explorations by its lips and tongue, but also the simultaneously tactile and kinesthetic explorations of chewing. Likewise, the grasping of a thing is not merely the tactile exploration of the sensual surface, but is often accompanied by spatial manipulations and banging against other available surfaces. What the child discovers in these interventions is the basic physical and causal dispositions of the thing; chewing can determine resilience and solidity, manipulations of any sort can determine extension, and motion teaches heft. Hammering (or proto-hammering, since this is not yet a specifically goal driven activity but an exploration) can reveal the powers and the limits not just of the "hammer" and the struck surface, but also of the infant itself; the different reactions of the "hammer" to different surfaces (the table, her face, a paper plate) slowly allows the thing to be placed in an ever enlarging web of things related by the threads of functional, causal, and simple physical interactive tendencies. The child comes to know quite soon that the bottle only delivers its contents in a certain attitude, when acted on in a particular manner; and it is by coming to place objects in causal-functional relations, to know the tendencies of a given thing (or given sort of thing), that the child can begin to *use* objects to bring about desired ends, coming to know better, and thereby extend, its own powers. "The picture of everydayness that emerges is of a holistic system of internal relations in which the 'ontological definition' of any entity is fully circumscribed by its actual place in an equipmental context."[120]

To know the entities of the workshop is to know them in terms of their causal tendencies, to be well enough familiar with them to manipulate them easily in the service of various ends, indeed, it is to know what ends they best serve.[121]

In the case of artifacts it is a crucial part of the picture that the primary functions of the tools of the workshop are socially defined; the encounter of the individual with the world is neither solipsistic nor idealistic, but is a process of the discovery of the order of things which pre-existed that individual. "Among humans, objects have standardized functions, and proper ways to use things develop among a group of beings who establish standard ends and communally shared ways of achieving those ends."[122]

Our encounter with the world, even the artifactual workshop, is a negotiation and not a definition of reality. Our understanding of the thing in terms of this function is not the idealistic covering up of the real, but the revelation of the given essence of the thing, which is just to say that here intervention provides the resources to support an epistemological *realism*.

For what, in the end, is claimed for embodied knowers is twofold: (1) they are in epistemic contact with the structured physical world *in virtue of* their agency, and the discernment of this structure is not just made possible by, but is in part constituted by practical interventions, and (2) intervention is a kind of knowing, its own mode of *epistemic* access to the world, generating a practical know-how which is a necessary condition for the "theoretical" or "conceptual" know-that. These two aspects of embodied knowing reinforce one another in various ways. In grasping the hammer, I have latched onto an actual individual part of the workshop environment, and in exploring with this object, in its gradual transition for me into a tool, I am discovering and discerning the real ontic and functional structure of that workshop. Knowing *that* it is a hammer is intimately bound up with knowing *how* to interact with, manipulate and use it: it is to know its place in the causal nexus which we with our active bodies explore. A being for whom such intervention was not possible, indeed, who was not in fact a largely competent navigator of the physical world, would not have acquired the empirical concepts by which we identify and sort the things of the world.[123]

But perhaps this claim is premature in two distinct ways. First, we have been dealing with examples in the context of the workshop, where a practical, functional ontological structure is the natural choice; it is not yet clear how applicable the example is to the world outside the workshop. And second, we have not dealt with explicitly in this chapter with language, and as Mark Okrent points out: "Absent language and given his premises, Heidegger's thesis (1) above, that there is no understanding-that without understanding-how is trivial, because the only possible evidence for beliefs must be given in the context of skillful, purposeful, action on the part of the being to whom the beliefs are ascribed."[124]

Without language, I can only display my knowledge that something is a hammer by displaying the know-how appropriate to that sort. But if I can speak,

I can simply say "There's the hammer." This raises the possibility that language introduces or allows a cognitive, observational encounter with the world which bypasses or transcends the practical.

Taking the second of these potential problems first, it must of course be acknowledged that once one has acquired a set of empirical-linguistic concepts, by, through and in terms of which observational data is organized into empirical knowledge, identifying a particular object as belonging in a given sort need not involve any actual intervention or manipulation. But the acquisition of the concepts involved in such identifications has, as a point of fact, involved interventions of various sorts. The very possibility of reference seems to ride on our capacity to individuate, which capacity depends in various ways on our physical interventions in the structural order. And the formation of our ideas of sorts seems to rely on physical interaction and exploration. All this is to say that language, too, is a "founded mode" of our being in the world.[125] Indeed, as a point of empirical fact, a great deal of our knowledge about the world and its objects (which knowledge, at least insofar as it allows us to apply given terms correctly, has a right to being called the "sense" of the empirical-linguistic concept in question) is interactional, functional, "knowledge-how." The question for epistemic theorists is how to analyze this; we have already, in various ways and in great detail, seen the epistemic and metaphysical consequences of taking intervention to be in itself an epistemically neutral activity, sometimes put to the service of producing sensually-available phenomena for conceptual processing. I hope to have laid the groundwork for a better analysis: the empirical concepts involved in structuring our visual-sensual-descriptive "representation" of the world play the particular synthetic/interpretive role they do because of (they get their particular epistemic role in virtue of), these practical interventions.

The first objection requires a more thorough answer, as it touches on subjects not yet extensively covered in previous chapters. We can elaborate that objection in the following ways: (1) It is indeed natural to suppose that our encounter with tools would be a bodily, practical, interactive engagement; but why should we suppose that this is an important component of our encounter with the natural world? After all, the most successful practice in producing knowledge about the natural world is the observational, theory-driven practice of modern science. (2) Perhaps the analysis of the nature of our ontic knowledge as a *functional* classification is appropriate for artifacts (which are, after all, designed for and defined by their function) but it is not clear that we can impose this sort of functional teleology on nature, nor can we suppose that nature is realistically intelligible in terms of human purposes.

Again taking the second question first, it is clear, I think, even without entering into the metaphysical debates about whether species possess actual distinct essences, and whether they are to be defined by "intrinsic" or "extrinsic" properties,[126] that the central defining features for natural kinds are their causal tendencies and functional dispositions. When a "real essence" is proposed for a species, it is by way of explaining the causal properties in virtue of which a given set of individuals is seen to form a natural sort; such metaphysical assertions must be made in the service of explaining or accounting for the causal structure of nature.[127] Thus the general strategy of identifying natural kinds with reference to their place in an extended web of causal and functional relations seems sound.

There is a different sort of worry, however: Heidegger, who argues "even as early as *Being and Time* that the praxical 'knowledge' implied in technology points to an interpretation of nature itself"[128] goes on to argue in "The Question Concerning Technology" that the effort to understand nature on the model of artifacts is in fact one source of our "covering up" of the natural order. So long as we see the world as existing *for us*, in the way that artifacts do, we are not seeing nature itself—for we are in effect imposing on the world our own system of needs and interests which prevents us from seeing the actual natural order; in its fullest and worst manifestations trees become for us nothing but pre-lumber, grass nothing but ornament, and animals nothing but food or early-stage petroleum products. What is so pernicious about this mistake is that it is not in any simple way a falsehood: trees *can* be lumber, we *do* have an aesthetic appreciation of the English-style lawn, and animals *are* important food sources (and when decomposed under the right circumstances form oil). There is a level of truth about nature which is revealed even in this most narrow-minded apprehension. Thus the criticism here cannot be that the actuality of our own interests must always prevent us from seeing nature aright, that the mechanics of our encounter with the world acts to keep nature at bay.[129] As already noted, we need to know a great deal about nature if we are to manipulate it to serve those interests. Rather, the mistake comes in the tacit assumption that the natural order, like the artifactual one, exists for us; in the absence of this assumption there is motivation to examine the relations of the entities of nature to each other, the ecological structure of the world as it is. Thus can we know the place of chlorophyll in the functional structure of *thymus vulgaris* (or perhaps discover the chemical origin of its wonderful taste and aroma); we can discern the intricate relations between ocean algae, CO_2 production, and cloud cover, and the role of mitochondria in cell metabolism. Of course, research *is* often motivated by the need or desire for greater control over a given phenomenon; knowledge is often not only gained through environmental manipulation, whatever its scale,[130] but also implies the capacity for further manipulation. But denying the knowledge-power relation gets us nowhere, politically or philosophically: here, as before, the difference between a world-hiding idealism and a real (even if narrow) realism is not an interested versus a dis-interested investigation, but whether and to what degree the information gathered from an investigation (however interest-driven) can be attributed to the world.

Thus even if we accept the principle that identifying the causal-functional tendencies and relations of things is the way to determine their natural sort, we must ask about the methods by which this is done: do we possess access to the world sufficient to allow us to make such determinations? Obviously, I think that we do, and in the last, brief section of this work I shall explain how and why.

4.62: Scientific Method as Embodied Access to the World

The objection that we are left with at this juncture is an objection from the context of a traditional picture of scientific method. According to this picture, we sensibly confront the world and, on the basis of some observed patterns in our sensations we "retroduce" an hypothesis about the structure of the world, which, if it were true, would explain much of the observed regularity. Whatever interference we attempt is done to produce more observable phenomenon, to see if they continue to fit discovered regularities, and to see if they conform to expectations deduced from our working structural model. As it was put by David Resnik, "In experimentation, we take advantage of the world's causal structure in order to produce, control, and observe phenomena."[131] On this model, the only epistemic conduit between us and the world is observation: we take advantage of the world's actual causal structure (whatever it is) to produce phenomena determined by the effect of our interventions on that real causal structure, to see if the results are the same as we predict from imagining (or computer modeling) the effect of the same intervention on a world structured like our hypothetical model.

A great deal of the intention of this work has been to insist that we are not, epistemically, where this model puts us. One important ally in this contention is Ian Hacking. Hacking is a recent, important champion of the idea that the primary producer of scientific knowledge is not theorizing and inferring, but experimenting and intervening: "I shall insist on the truism that experimenting is not stating or reporting but doing—and not doing things with words."[132] For Hacking, experimentation has a life of its own, and epistemic import which goes beyond whatever service it may render theorists:

> Experimental work provides the strongest evidence for scientific realism. This is not because we test hypotheses about entities. It is because entities that in principle cannot be "observed" are regularly manipulated to produce new phenomena and to investigate other aspects of nature. They are tools, instruments not for thinking, but for doing.[133]

We can, I think, read this passage to be in accord with Heidegger's workshop metaphor: our encounter with the world, even our scientific encounter, is not well described (solely) as inference from observation. To say this is to forget that what we are doing in the interest of that observation is *manipulating*, using entities as tools, taking advantage of known[134] causal tendencies to interact with them.

Hence from the very beginning people were less testing the existence of electrons than interacting with them. The more we come to understand the causal powers of electrons, the more we can build devices that achieve well-understood effects in other parts of nature. By the time that we can use the electron to manipulate other parts of nature in a systematic way, the electron has ceased to be something hypothetical, something inferred.[135]

It would seem silly to doubt the existence of the tool one is using to interfere with and effect the world, but, and this is crucial for understanding Hacking's point here, the reason it would be silly is *not* because in the course of the experimental interference, we produce the observable phenomena we expect, thus confirming our model of the structure of electrons. The reason it would be silly is the same reason we normally would take it to be silly (or even mad) for the carpenter to doubt his hammer's reality: because we are *using* it. "Prescott, et. al. don't explain phenonema with electrons. They know how to use them."[136] Our bodily interference in the world, our manipulation of its entities, is itself a kind of knowledge (not reducible to sense-data) which we would simply be ignoring to use something and simultaneously to deny its existence.

The failure to understand this leads to two kinds of bad arguments against Hacking's position: that his is just another version of the "argument from success," and that experimentation is not, in fact, "theory free."[137] The "argument from success" or "inference to the best explanation" is just the traditional model of scientific explanation we sketched above, while the latter, as we might expect, involves the charge that experimentation is not as independent of theory as Hacking insists, and thus that experimental observation fails to escape the dilemmas discussed in the beginning of this work (and Hacking's). Take the following passage from Reiner and Pierson, in which the two objections are somewhat mixed:

> If we have access to certain unobservable entities through the laboratory skills by means of which we manipulate them, and if our confidence in these skills does not depend in any serious way on any scientific theory being true . . . then it looks as if we should have grounds for rational conviction that these entities are real. Things are not, however, this simple. Laboratory skills do not give us access to otherwise unobservable entities, but only to certain observable interactions in the apparatus. Only by IBE [Inference to the Best Explanation] can we come to believe that these observable signs indicate the presence of causal interactions . . . Moreover, only by an *additional* IBE can we take the evidence as warranting belief in the existence of exactly one kind of entity, rather than two kinds or a thousand kinds.[138]

Note that the critique is predicated on denying precisely what in their assessment they grant Hacking: that interaction gives *access* to entities.[139] Of course, if we have no such access, if our epistemic relation to the world is exhausted by perception, then we are in the very boat which Reiner and Pierson describe. Hacking is well aware of this: it is what he wrote his book to avoid. Although I cannot claim that Hacking was always clear about his epistemic position (so that I cannot be sure that he would agree to being placed so squarely in the context of the current inquiry) it still seems a poor, and invalid, argumentative method to assume, in the name of your interlocutor, a position he is bound by his own conclusions to deny. Any critique of the interactionist paradigm must begin by taking seriously the claim that interaction is *epistemic* access to the world.

The second sort of misguided objection focuses more on Hacking's claims that experimenting is independent of theorizing. Hacking certainly claims that the epistemic results of experimental interaction with entities do not depend on the truth of theories, so that, for instance, if we have good experimental evidence for the existence (and characterization) of a given entity, then this entity will survive theoretical paradigm shifts. Only someone who believed that the structure and organization of our theoretical and conceptual hierarchy *defined* reality would deny this thesis, so Hacking seems on pretty firm ground here. The question is, does Hacking have to assert further that experiments are independent of theories in that they are not guided by them? One can easily see the context within which such a concern would arise: if we accept the traditional model of science, all experimentation is driven by theoretical expectations, designed to confirm (or refute) a given structural hypothesis, and the phenomena which experimental intervention produces get their significance only by being placed in this context, interpreted in light of current hypotheses. If we can deny that experiments are all theory driven in this way, we might have some evidence that their results were also theory independent (did not depend for their meaning or significance on a given hypothesis, but instead in some sense revealed the world as it is). But this worry only arises in the context of the acceptance of the traditional model, and for this reason it is a bit worrying that Hacking spends some rhetorical energy identifying experiments done without theoretical guidance, or whose significance was not in fact a matter of interpretation by current hypothetical standards.[140] Not only does this make it seem as if he were still tacitly maintaining the traditional model, but it is much too easy to counter Hacking's examples with others, showing the general reliance of experimentation and intervention on goal-driven intention and theory.

A better, and more fruitful strategy, is simply to acknowledge the fact that, although "experimentation is a form of intervention . . . it is intervention strongly guided by representation."[141] *Of course* our ideas, beliefs, and "representations" of the world guide our interaction with it; what we know about the world must guide our activity if we are to have any sort of success. But it is equally true that our interaction and manipulation of the world guides our representation of it; and although the traditional model "saves" this phenomenon of interaction contributing to knowledge, it does so at the expense of realism.[142]

The analyses of the current work do not foolishly depend on denying the importance of theories to scientific knowledge, the role of inference and retroduction in the formation of those theories, or the role of theories in guiding experimentation.[143] Theories often give us great insight about how to manipulate and at what juncture to interfere. But I do want to insist, with Hacking, that entity realism need not be rooted in "proving" the truth of these immensely helpful theories; for we have a mode of epistemic access to the world the results of which are not mediated by the theories which may guide interventions, for the results are not only observations, but coping skills. These coping skills, the interactive comfort which allows one to negotiate a given environment, constitute a form of knowledge to the service of which our representations are bent and by which our representations are guided. (see, e.g., sec. 4.3)

Let us consider in more detail an actual, albeit simple, laboratory situation. Following Hacking's paradigm case of microscope use, we can imagine the following: Coming to a microscope for the first time, a student peers through the eyepiece. Seeing nothing (i.e., seeing only unbroken light) she *fiddles* with the focus knobs. After some movement, a pattern of dark and light, probably a very complicated pattern, emerges. The student *blinks* (washing her lenses) but the image remains, the student *wipes* the lenses of the scope, but the image remains. The student *moves* the specimen slide, and the image moves, too. Now the student takes a small probe and *moves* it under the scope, onto the slide (a large, very dark shape enters the view, at which point two things become apparent: first that manipulating the image area actually alters the image, and second that our visual representations and our motor space are linked up in a very specific way which no longer applies in this new representational space. The probe visually moves in directions other than we expect. This takes some getting used to.) Taking the probe we *push* on the edge of the pattern and a surprising thing happens: some of the pattern moves while some remains, or scatters in directions other than the direction of the exerted force. Repeated *manipulations* of this sort leave us with a pattern which is relatively stable, and tends to react in predictable ways to the *interference* of the probe. To argue that the natural conclusion of this sequence, that we had identified an actual, microscopic individual, is an instance of "inference to the best explanation" ignores the fact that we were not in the position of investigating a visually clear phenomenon with interventions designed to produce more phenomena; the intervention helped *define* the visual phenomena, in terms of its reactions to the interventions, as corresponding to an individual. As Hacking puts it, "we don't just peer, we interfere."[144]

Now let us imagine that what we had thereby individuated was a living cell. There are, of course, distinct parts to the cell, and we could follow the same steps above, perhaps with a higher power lens, to identify "visually" different internal structures of the cell. But what do these structures do? What are they for? What *kinds* of things make up a cell? To answer these questions it is a sure bet that more interference and manipulation is required. We might notice, for example, that the cell tends to move away from bright light, or moves more slowly in certain kinds of media; and we might suppose that some of the individual structures we identify within the cell may be responsible for these behaviors. This, of course, is an hypothesis, the testing of which requires very directed sorts of interventions: we might destroy a specific part of the cell (what we suppose to be its "eye") and determine if it is still antiphototropic. Likewise we might put a number of cells in what we know to be a nutrient rich medium, and, by destroying select parts of the cells, determine which parts are responsible for digestion, or removing waste matter. As Hacking writes, regarding genetic microscopy:

> I must repeat that just as in large scale vision, the actual images or micrographs are only one small part of the confidence in reality. In a recent lecture the molecular biologist G. S. Stent recalled that in the late forties *Life* magazine had a full color micrograph, labeled, excitedly, 'the first photograph of the gene' (March 17 1947). Given the theory, or lack of theory, of the gene at that time, said Stent, the title did not make any sense. Only a greater understanding of what a gene is can bring the conviction of greater understanding of what the micrograph shows. We become convinced of the reality of bands and interbands on chromosomes not just because we see them, but because we formulate conceptions of what they do, what they are for. But in this respect too, microscopic and macroscopic vision is not different: a Laplander in the Congo won't see much in the bizarre new environment until he starts to get some idea what is in the jungle.[145]

In the above examples, supposition, hypothesis and the production of visual phenomena have their place, but so too do interaction and manipulation. An indispensable part of having an "idea of what is in the jungle" or under the microscope, which allows one to see properly, is determining the functional place and interactive tendencies of the things one finds there. Sorts are discovered, defined, and seen in terms of their effective place in a web of interactions and processes, and we can only determine the characteristics of this place by intervening in, and sometimes disrupting, the natural processes themselves.

The epistemic importance of interaction and intervention is taken one step further in Hacking's paradigm case of electron microscopy. What rightly strikes Hacking as remarkable here is not the amount of interaction which occurs with the specimen to be explored, but the degree of confidence and comfort one must have in one's interactions with the electrons. For clearly one has entered into a relation with the electron that is far more intimate even than one's relation with the observed, poked and prodded cells in light microscopy; a stream of electrons has taken the place not of the specimen, but of the *probe*, the *tool* of interference and interaction wielded by the microscope operator. This, perhaps, is what is so striking about the famous IBM image, constructed by carefully placing individual atoms in the form of the company logo. Of course it indicates that we know a great deal about the behavior of atoms, enough to get them to line up in an extremely unnatural way, but more than this it indicates a level of comfort with the technological tools of the laboratory, such that the phenomena of which they take advantage are no longer known in terms of any inferential knowledge-that, but in terms of an interactive knowledge-how. Just as our epistemic encounter with a hammer qua hammer is poorly described in terms of a sensing subject hypothesizing the properties of the sensed object, and interacting with the object in the hopes of confirming the hypothesis (with whatever confidence eventually gained in our understanding of the hammer being ascribed to the continued confirmation of our phenomenal expectations) so, too, once the electron has been so thoroughly harnessed as to be used continually and successfully as a tool, it would be a mistake to continue to describe our epistemic relation with electrons in terms of inference and hypothesis; knowing is not just seeing, but also coping.[146] It is just one more version of this same mistake to treat our interactive comfort as a by-product of some long process of purely sensual information gathering, to treat our coping as if it were grounded in seeing; interaction is a mode of epistemic access, producing its own knowledge, from the first epistemic encounter.   When a sort of object, once hypothetical, becomes available to us as a tool, this indicates that our interactive access to the things must have grown steadily, providing us with knowledge-what (understanding the characteristics of their causal-functional place) in virtue of, and simultaneously with, knowledge-how. In the successful scientific endeavor, hypothesis gives way slowly not (just) to a catalog of observed phenomena, but to the slow inclusion of the entity in that extended web of functionally related entities with which we have learned to cope. This is why we should take the central symbol of the success of western science to be not its extension of our senses, but its extension of our bodies. Like Merleau-Ponty's blind man who can "see" (determine the physical layout of, and thus cope with, his environment) with his cane, we extend our interactive capabilities often without extending our sensual capabilities; the cane has no nerve endings, but, as a tool of interaction and interference, can reveal for us the structure of the physical world.

One approaches the world by interacting with it, inserting oneself into the functional order, coming gradually to identify objects in terms of their place in the network of causal and equipmental contexts which we negotiate each day: chairs are for sitting, cups for drinking, and beds for sleeping. Likewise, interactions with the microscopic things of the world reveal for us their place in the extended causal networks with which scientists are familiar, and accepting their reality involves fitting them into that network which is ultimately continuous with the everyday. Indeed, the suggestions I have made here are simple, even if their consequences prove difficult to trace. First I suggest that our primary epistemic relation to the world is not one in which an "external" world is present to us in the form of sensations from which we are meant to infer the structure of that world. Rather, we come to know our world through the activity of being in the world. Second, I have suggested that the world and its objects might usefully be characterized (metaphysically) in functional/ecological terms. To accept both of these suggestions equips one to see beyond simple skepticisms, for on this model there need be no necessary, unbridgeable gap between knowledge and reality, whether in science or in everyday life.

Notes

1.   Michael Ayers, *Locke* (London: Routledge, 1991), 14.
2.   John Locke, *An Essay Concerning Human Understanding*, (New York: Dover, 1959), IV, iv, 4.
3.   Locke, IV, xi, 2.
4.   It does not, however, follow that if we do make this judgement that the experience is knowledge. The judgement that there is an external origin of a sensation (that a sensation is, as we might say, a perception) does not guarantee its veridicality. See my comments on Descartes in chapter 1.
5.   Ayers, *Locke*, 155. Ayers notes: "For Descartes, as it seems, the representative content of a sensory idea is determined by how the mind relates a datum to a supposed cause, an act which can be done well or ill, perversely, or naturally, or rationally. But for Locke, once the mind uses the datum as a sign (and that seems to be assumed as inevitable), what the idea signifies is itself in effect a datum too. A simple natural sign can only signify its normal cause, whatever that may be." Ayers, *Locke*, 157.
6.   Locke, II, xxiii, 2.
7.   Locke, II, xxiii, 10.
8.   Locke, III, iii, 17.
9.   There is, of course, a sense in which I am reasserting Locke's distinction between primary and secondary qualities, but only epistemically, for in so doing I am questioning Locke's epistemic empiricism/sensualism according to which the same mode of epistemic access accounts for our knowledge of these different aspects of the material world. See *Colour Vision*, by Evan Thompson, for an excellent discussion of the limits and pitfalls of this *metaphysical* distinction.
10.   I here gloss over the distinction between primary and secondary qualities because both have practical import, and it is clear for Locke that we come to *know* these properties in the same manner. It is this part of Locke's epistemology which made it so easy for later empiricists to deny our knowledge (and therefore the existence) of matter as such. By collapsing primary and secondary qualities, the skeptical empiricist argued that we could know nothing of what the world possessed "in itself." See in this regard Ayers' justified chastisement of Locke for adopting a simple sensualist position with regard to our knowledge of matter (Ayers, *Locke*, 182).

11. Locke, II, xxiii, 32. Also consider: "We are then quite out of the way, when we think that things contain *within themselves* the qualities that appear to us in them; and we in vain search for that constitution within the body of a fly or an elephant, upon which depend those qualities and powers we observe in them." (Locke, IV, vi, 11).
12. Locke, III, iii, 18.
13. Locke, III, iii, 13.
14. Locke, III, iii, 12,15.
15. Locke, III, vi, 33.
16. Locke, III, vi, 50.
17. But it is important to realize that, even granting our limited knowledge and the bizarre metaphysical conjectures above, we would not thereby be "mistaken" in our classifications; when we determine that a given nominal essence is appropriate, then it is, ex hypothesi, appropriate to class the parcel of matter under that nominal essence *regardless of its real essence.* (Locke, IV, iv, 11).
18. Locke, IV, vi, 11.
19. Locke, IV, vi, 11. It seems perverse to me to assert that the knowledge "forced" upon us by the necessity of breathing has come by *sensual* means. What sort of quale might these be that carry the knowledge of our need to inhale?
20. Locke, III, vi, 9,27,32. And this is just Locke's point: our schemes of classification, being epistemically rather than metaphysically driven, must, and do proceed with no regard to the natural order, per se.

Obviously, however, the way substances affect us is part of the natural order, as are many of our interests, e.g., eating and staying warm. But Locke is saying that at the level at which these causal necessities are interests, i.e., objects of our conscious awareness, then they are ideas, and therefore part of the phenomenological, and not the causal-physical order. As noted, they do reveal something true about that causal order, *but not in purely causal terms.* Our sensual need indicates to us a physical origin (as our reception of simple ideas indicates to us an outside causal force), but our knowledge goes no further than the realization *that* there is something without us with the power to produce a given idea in us.

Just as obvious is the fact that our classification can never be purely "interest" driven, either; if we have an interest in hammering, then we must find a substance which possesses the requisite weight and hardness for the task. Still, for Locke *how* it has the power to feel hard (and to affect things commensurate with our expectations of how a hard thing would affect them) is a causal mystery forever hidden from us because we are epistemically limited to the sensual/phenomenological order.

21. Locke, IV, vi, 4.
22. As opposed to, say, limitations in our understanding, or the unlimited metaphysical complexity and randomness in the natural order.
23. Locke, IV, xii, 10. "I would gladly meet with one general affirmation concerning any quality of gold, that any one can certainly know is true [of "real" gold]. It will, no doubt, be perfectly objected, Is not this an universal proposition, *All gold is malleable?* To which I answer, it is a very certain proposition, if malleableness be a part of the complex idea the word gold stands for. But then there is nothing affirmed of gold, but that that sound stands for an idea in which malleableness is contained." (Locke, IV, vi, 9).

It is certainly worth noting the way in which Locke's sensible empiricism limits and mars his account of knowing. Surely no sensation (or even definable set of sensations) corresponds to the complex property "malleable." This is rather a way of characterizing a certain substance's typical reaction to given physical interactions or pressures on it: the parcel of matter will tend to deform with the application of pressure.

24. See Michael Friedman's *Kant and the Exact Sciences* for an illuminating account of Kant's attempt to preserve our epistemic access to the physical, causal structure of the universe, our awareness of our real interaction with substances, against the analyses of Leibniz.

25. "Solidity is in itself a property of matter like extension and motion. . . . According to the doctrine of simple ideas, more neither can nor need be said: 'If anyone ask me, what this Solidity is, I send him to his senses to inform him: Let him put a flint, or a Foot-ball between his hands; and then endeavor to join them, and he will know.' The claim was, then, that the simple idea of solidity is an indefinable *quale* which represents and 'resembles' that property of matter in virtue of which objects press upon, and resist the pressure of, other objects including ourselves. The whole model crumbled before the onslaught of Hume, but Hume characteristically rejected, not the deeply suspect notion of the simple sensory *quale*, but the possibility that this item, a tactual feeling, should be *like* solidity, if solidity is that in the object which constitutes its impenetrability. The alternative path, just as destructive and more illuminating, is to find an acceptable sense in which our sensations of solidity do represent an object as it is, but to reject the conception of the simple, blank, *quale*." (Ayers, *Locke*, 182).

Ayers continues to suggest the outline of just such an account, as both preservation and critique of Locke's empiricism. My own response, of course, is to question whether, as Ayers says, "solidity is after all a sensible quality." My analysis of Locke's own example above is to cast our solidity (and the experience of solidity) in kinaesthetic terms, consistent and coherent with the tactile, but representing a different conduit for knowledge.

26. This looks rather similar to John McDowell's formulation, which could indicate that, contrary to the reading of his position I gave in section 3.3, he is in fact very close to Locke. However, Locke's unique position regarding our sensory reception of reality is possible *only* because of his distinction between primary and secondary qualities, *and* his insistence that there is neither Cartesian-style judgement nor conceptual synthesis which plays a role in our reception of conceptual content. These positions are not readily available to post-Humean, post-Davidsonian McDowell; whatever his respect for Locke's commonsense conclusions, he requires a different, non-Lockean perceptual psychology if he wishes to support them.

27. It is not at all clear to me that the causal connections insisted on by so many epistemic theorists carry more epistemic value than this: because of my causal reception of the world, I can know that there is a world. But after accepting Hume's critique of Locke, we cannot claim that we know how the existing world is (except "to us"). The most sophisticated contemporary epistemic systems are too often just variations on this theme, thought to be a truism: that we only know how the world is "to us" (where this is meant to *imply* that we do not, or cannot, know how the world *is*—that "to us" always necessarily implies an attenuation of realism or objectivity).

28. Note that I do not argue here that Locke's model was *successful* in arguing for the immediacy of our awareness of the physical structure of the world, nor can I argue that Locke's model *in fact* preserves that epistemic openness to individuals which he denies us in the case of sorts. I can only claim that my treatment reproduces Locke's intentions; it must nevertheless be admitted that Hume attacked Locke at a point of genuine vulnerability.

29. If I am right to suppose (as I argued in chapter 3) that the capacity to make such attributions is a condition of realism, then it is not at all surprising that Locke should not be a realist about kinds. Given this it is also not surprising that Locke should be pessimistic about Natural Science: empirical science is, first and foremost, the investigations of the rules governing the interaction of individuals, *precisely insofar as those individuals are representative of general types*. But if the general types which we know are in fact defined by us, there is nothing for natural science to discover: their interactions will be a consequence of our own delimitation of complex ideas. If there is any discovery to be made, it is of the relations of ideas. Thus the science of Logic occupies for Locke the place normally reserved for a science of Nature. This is a consequence Locke embraces: "General and certain truths are only founded in the habitudes and relations of abstract ideas." (Locke, IV, xii, 7).

30. Locke, III, vi, 33.

31. Stephen P. Schwartz, ed. *Naming, Necessity and Natural Kinds* (Ithaca, NY: Cornell University Press, 1977), 15. "The central features of what is here meant by a traditional theory of meaning are the following. (1) Each meaningful term has some meaning, concept, intension, or cluster of features associated with it. It is this meaning that is known or present to the mind when the term is understood. (2) The meaning determines the extension in the sense that something is in the extension of the term if and only if it has the characteristics included in the meaning, concept, intension, or, in the case of the cluster theory, enough of the features. In many contemporary versions, the meaning or concept of the term may include only observable criteria for the application of the term. (3) Analytic truths are based on the meanings of terms. If P is a property in the concept of T, then the statement 'all T's are P' is true by definition."

Of course, Locke, by denying that sortal terms refer at all, is not an adherent of the traditional view.

32. I stand by this statement so far as it goes: but see my important qualifications below, for the remainder of the chapter. Accurate reference can be attributed to our *possession* of a description, although not autonomously to the power of the description alone.

33. Deeper reasons for abandoning phenomenological resemblance theories, and, indeed, any theory of the aboutness of mental states which rests primarily on our sensual detection of properties, can be found in Kathleen Akins, "Of Sensory Systems and the 'Aboutness' of Mental States," *Journal of Philosophy* 93, no. 7: 337-372.

34. This result has nothing to do with the fact that I have cast the descriptions phenomenologically. I could just as well have said that I do not thereby refer to the one which *really* "has the power to cause blue in me" (in this case the one that looks like some other color). And I certainly do not refer to the object-contact lens combination (which in combination has the power to cause blue when we look at a green object). I refer to the one that *looks* blue; but not, I think, in virtue of the content of the idea, but rather because that is the one I intend. The question becomes, in virtue of what does intention guide reference?

35. Which is just to say that Puntam, Kripke, Donnellan and many others are surely right to be critical of what they call the "traditional" theory of reference. Of course, such a theory is not Locke's.

36. And it *still* wouldn't follow that *descriptions* were the agents of reference here. To paraphrase Austin: it is *we* who refer with words.

37.   Note that this only illuminates the *general* relations between epistemic and referential capabilities. I am claiming, in general, that it will only be possible to *refer* to a given kind of thing (individual, sort, etc.) if it is possible to *know* that kind of thing. I am *not* arguing that one can refer to only those things with which one has currently (or even has had in the past) an actual epistemic link. Determining the specific epistemic capacities which would be drawn into operation in any actual instance of reference to a given particular or sort is a different, and more complicated, undertaking. There are many varieties of reference, and many complex considerations for each variety. (Such as the different kinds and degrees of reliance on an epistemic community in reference and language use.) For a careful discussion of these more particular issues, see Gareth Evans *The Varieties of Reference*.
38.   Thus, given the criterion discussed in chapter 3, Locke must adopt an anti-realist account of kinds.
39.   That is, in just the position in which Hume casts the human knower with his critique of Locke, a position from which, if I am right, neither Kant nor the contemporary inheritors of the notion of conceptually-synthesized experience as the basis of knowledge have the means to escape.
      Putnam reaches the same conclusion in his defense of internalism, and so we place the knower of "contemporary" theory in a position similar to that of Descartes' confused dreamer (from chapter 1).
40.   Hilary Putnam, *Reason, Truth and History* (Cambridge: Cambridge University Press, 1981), 62-4.
41.   Putnam, *Reason, Truth and History*, 52.
42.   Whether it is also, in fact, "mind-projected" in the fully Kantian sense is less clear. He says, of course, that "the mind and the world jointly make up the mind and the world," and this means at least that (1) we understand the world in terms of our conceptual-linguistic cognitive architecture, and (2) there is no "given" in experience to which this architecture can be unproblematically attributed.
43.   Putnam, *Reason, Truth and History*, 54.
44.   Descartes' definition of an object from the *Principles* is one example of both the tendency and its consequences.
45.   Unless we take our conceptual synthesis of perceptual information to *be* individuation, *defining* the individuals and sorts of the world. But this is just the disease I want to avoid.
46.   But to treat these boundaries as artifacts of the representational matrix is to treat them as divisions for which we are responsible, and without convincing reasons for treating these artificial divisions as grounded natural divisions, we cannot be realists. Indeed, accepting as artificial the principles of division necessitates a turn to mass-artifacts like society as a way to avoid epistemological solipsism. What emerges from this turn is a politicized epistemology based on a kind of solidarity of enforced agreement on our culturally-rooted principles of division and conceptualization. This is why Richard Rorty's metaphysical epistemology has been reduced to pleas for liberalism; in this context philosophy becomes a kind of activism as we all fight for control of the principles of division which determine the conceptual structure of our society.
47.   I pick up this term from Heidegger. It seems singularly appropriate for the purpose: we say of an animal or human that it is autonomous (self-moving), so we can say of the material particular that it is self-standing, and this is meant to have the same connotations of independence which we attach to autonomy. In Heidegger's resonant phrase: "The thing conjoins itself out of a world." Martin Heidegger, "The Thing," in *Poetry, Language,*

Thought, translated by Albert Hofstadter (New York: Harper & Row, 1971), 165-182.
48. Heidegger is by now well-known for his insight that environmental mis-use can be traced back to the denial that the ecosystem has any inherent structure, any structured existence besides subject-projected use-values. As both trees and humans succumb to this covering-up, so too do well grounded moral limits on acceptable comportment towards the objects in question. (See Martin Heidegger, "The Question Concerning Technology.")
49. Putnam, *Reason, Truth and History*, 53.
50. Imagine a table with two objects, one falling under the nominal essence of gold, while the other is apparently not gold. But imagine that they share the same real essence. In such a case, the person who asked "Would you bring me the gold from that table?" would have referred to both items on the table, *even though he didn't realize it*. For the sake of clarity, I will call those objects which fell within the language user's realization (within the acknowledged scope of his request) his *intention* (as in: he intended to refer to X). Likewise, if the situation were reversed, so that both objects had the nominal, but only one the real essence of gold, the speaker would have referred to only one, even though he would be quite puzzled (or annoyed) to have had both objects brought him in the first case, or only one brought in the second. We can suppose that were he to *discover* the cases to be as we have described, he might say he was mistaken about how many gold objects were on the table, but it surely does not follow that he had referred to objects other than those he intended. (Actually, it is not necessarily clear that being presented with the above scenarios would change one's opinion on the extension of given terms: it is always an option to respond "that is not what I mean by 'gold.' I meant just what I intended." And in any case, it is extremely difficult to predict what we "would say" if confronted with the reality of the scenarios sketched above. It would, after all, entail coming to have an *epistemic* relationship to real essences which we have heretofore, *ex hypothesi*, lacked.)
     To suppose that reference *could* be an autonomous function of words, potentially disparate from intention, would have the odd result that we would be under constant danger of uttering falsehoods, (e.g., "There is only one gold item on the table.") and acting inappropriately in response to requests (e.g., Bringing one, but not both, items from the table) without ever having the suspicion that this was the case, nor having the means to confirm or lay to rest whatever suspicion we might develop (e.g., after taking a philosophy course). That is, we would be in such danger if we thought, naturally, that reference was an important factor in determining truth value; but to deny this would remove much, if not all, of the motivation for having a theory of reference.
51. Putnam, *Reason, Truth and History*, 52.
52. I certainly recognize the truth of Putnam's corrective to traditional accounts of reference on at least these two points: (1) "ideas" or sets of descriptions do not autonomously guide reference; (2) often a given term refers in the absence of the utterer's ability to pick out or recognize the referent, in virtue of her membership in a linguistic community which contains "experts" who can properly apply the term.
     But see the section 2.31 for thoughts on the limits of these insights. It should be noted in particular that it is not a consequence of corrective (2) that a given term can refer if an entire epistemic community is in a state of ignorance.
53. At least not without violating Russell's Principle (see below), which is, I think, an excellent, simple tool for guiding our understanding of the relations which must hold between our linguistic, conceptual and epistemic systems.

54. I do not here argue that the analyses of chapter 3 invalidate Ayers' account. But his account does have to answer the sorts of concerns raised in section 2.3.

55. Another aspect of our sensory field, its temporal development, although not a concern of the present work nevertheless deserves mention here. It is of course essential to our knowledge that it be temporally sensitive, by which I mean to include sensitivity not just to the changes in the spatial structure of the sensory field over time, but also to the persistence of spatially placed solids, and to the impermanence of more ephemeral phenomena. It is my feeling that bodily activity, because it is itself a temporally structured phenomenon, is an ideal candidate to account for the temporality of our sensory-field and the empirical knowledge it represents.

56. On the one hand, if we allow the colloquial use of "knowing which" to stand, the principle is probably false, as one can see an object and make judgements about it without knowing, say, what the object is. Yet if any sort of answer to the question "Which object are you thinking of?" will do, then the principle is trivial, since, as Evans puts it, "anyone who is willing to ascribe to a subject the thought that *a* is *F* in the first place will also be prepared to ascribe to him the thought, and presumably the knowledge, that it is *a* that he is thinking about." Gareth Evans, *The Varieties of Reference*, (Oxford: Oxford University Press, 1982), 89.

57. Evans, *The Varieties of Reference*, 89.

58. What is at stake here is an understanding of our capacity to think, speak and make judgements about particulars. The continuity of this exploration with that of chapter 2 should be clear: to give an account of this individuating knowledge is precisely to take up the project of ontology deferred by Frege.

59. This presumes that spatial position is not part of the "description" of an object. It is open for the imagined theorist to argue that our capacity to fix on such a particular *is* based on description, because location is a descriptive quality. Such a claim would involve an account of the perceptual quality of locations (and, presumably, an analysis of space in terms of our sensory field). Without constructing or recounting such theories in order to examine them in some detail, I can only say that it seems to me more natural to analyze our experience of space in behavioral terms.

60. This is another reason not to conflate the capacity to discriminate with the capacity to individuate: for surely the capacity to discriminate *between* things is based precisely on sensitivity to the content of the information gathered from them (gathered *in* objects).

It is important to insist that considerations regarding the content of information cannot be entirely set aside. One cannot be said to have a link with an object if one is acquiring blatantly incorrect information about it; one would not be *seeing* a chair which had all the appearances of a Bengal tiger. This introduces accuracy as a limit-case on the possibility of an information link; although the fix does not *consist* in accurate information, it cannot *persist* in its absence. To say that the link does not consist in accurate information is to emphasize the fact that the accuracy here is a necessary but not sufficient condition for that link; as noted, it fails to explain the *particularity* of the link. It is only on the assumption of the "bad old philosophy of mind," according to which the referent of a name (or idea) was whatever object best matched the descriptions associated with that name, that accuracy explains particularity. See Gareth Evans, "The Causal Theory of Names," in *The Philosophy of Language,* edited by A. P. Martinich (Oxford: Oxford University Press, 1985).

61. Here, and in what follows, I return to the more colloquial use of "object." I find the constant use of "thing" too hard on the mental ears!

62. Although of course the perceptual-information field is spatially structured, I claim that this structure originates in our embodied access to the world.

63. As above, it is easy to imagine failing to know "where the sound is coming from" in the sense of failing to know the origin of the sound in descriptive (as opposed to indexical) terms. Thus one may fail to know that the sound came from some particular factory in the distance, or one may fail to know because of an echo effect that a sound is coming from around the corner. But in each case one may place the sound in more or less vague directional terms: "over there."

64. I suspect that hearing is such that it cannot seem that a sound is coming from *nowhere*. Sound cannot be informationally utopic. It is not clear what is to be said about a ringing in one's ears or other auditory hallucination. I would say that such sounds always *seem* to be coming from someplace in particular, or from everywhere, but are *in fact* coming from nowhere (are not "sounds" at all, or this is not an instance of "hearing"). But if someone argued that one can experience such "sounds" as coming from nowhere, and further insisted on calling this experience hearing, I think this experience would nevertheless not count as the gathering of any *information*. In every case where it seems to a subject that they are gathering information, the experience includes some sense of the "placedness" of that information. It would seem that the sound originated in some (more or less vague, as per the circumstances) place.

65. Evans, *The Varieties of Reference*, 154.

66. Evans, *The Varieties of Reference*, 153-4.

67. Evans, *The Varieties of Reference*, 156.

68. Evans, *The Varieties of Reference*, 157.

69. I should note right off that, as Evans puts it, knowledge of a location in ego-centric space is not knowledge of a special kind of space, but a special kind of knowledge of space. "It is perfectly consistent with the *sense* I have assigned to this [ego-centric] vocabulary that its terms should *refer* to points in a public, three-dimensional space." Evans, *The Varieties of Reference*, 157.

70. The connection between space and behavior is hardly novel. Charles Taylor writes: "take the up-down directionality of the [perceptual] field. What is it based on? Up and down are not simply related to my body—up is not where my head is and down where my feet are. For I can be lying down, or bending over, or upside down; and in all these cases 'up' in my field is not the direction of my head. Nor are up and down defined by certain paradigm objects in the field, such as the earth and sky: the earth can slope for instance . . . Rather, up and down are related to how one would move and act in the field." Charles Taylor, "The Validity of Transcendental Arguments," *Proceedings of the Aristotelian Society* 79, (1978): 154.

71. One *can* know a location objectively and fail to know ego-centrically, e.g., "in Washington D.C.," but not in the case where the knowledge is being gathered via a perception of the object or location in question, and not generally when the location-information is to be effective in determining behavior.

72. Evans, *The Varieties of Reference*, 169. Note that the example exhibits the independence of spatial knowledge from perceptual information. One needn't be receiving information from a location to place it in ego-centric space, to fix epistemically on that

location. The behavioral space here precedes the informational field spatially experienced, although this is *not* to say that some transcendent behavioral field must be in place, or that we must have a settled and complete sense of our behavioral space before any perception is possible.

73. Evans, *The Varieties of Reference*, 161-2.

74. This, of course, is a point fundamental to the work of Merleau-Ponty. The organs of perception are physically and spatially active, and this activity—say the movement of the eyes in the course of a scanning-examination—is fundamental to placing the information, and is thus fundamental to the interpreted *significance* of that information.

75. Evans, *The Varieties of Reference*, 160.

76. Of course, for the place to be fixed in a behavioral space does not imply that it is accessible only by a fixed series of actions. One can alter one's position in relation to the place/object in question (and this of course alters the dispositional significance of any information gathered from that place; being further away one is less inclined to jump should a fire begin "there") without thereby losing one's cognitive fix on the location, without treating it as a "different" spot. That is, the location which I now know in terms of reaching my arm out straight to the left does not itself move 90° clockwise when my body does. The same spot now has a different behavioral significance, the particular character of my behavioral disposition regarding the place has altered, yet my knowledge of the place is still cast in ego-centric, comportmental terms.

77. It is important to note just how unmediated this connection is. We cannot imagine that, for instance, appropriate behavior is calculated from perceptual information (information whose content is not at root behavioral) for in such a case we would have to imagine a sort of conversion process which allowed the non-behaviorally cast perceptual information to be converted into behaviorally relevant data. Evans writes: "When we hear a sound as coming from a certain direction, we do not have to think or calculate which way to turn our heads (say) in order to look for the source of the sound. If we did have to do so, then it ought to be possible for two people to hear a sound as coming from the same direction (as 'having the same position in the auditory field'), and yet to be disposed to do quite different things in reacting to the sound, because of differences in their calculations." Evans, *The Varieties of Reference*, 155. Were two people meaning to indicate the same object to point in different directions, we would count the failure not as one of calculation (while allowing that both subjects had identical knowledge of the location of the perceived object), but as a failure in knowledge of location.

78. Evans, *The Varieties of Reference*, 163.

79. Evans, *The Varieties of Reference*, 172.

80. It is worth continually stressing that these two are not isolated from one another: in a sense, the behavioral field of ego-centered space plus the informational field of sensation together compose the experienced spatial world. I ought also to acknowledge the affinity of this account of space and information to Kant's. Of course, I hope to avoid Kant's idealism.

81. Our capacity to place non-material objects (rainbows and such) depends primarily on our sorting of perceptual information in behavioral space, and more specific placements are parasitic on our capacity to place material objects; we experience rainbows within a field of spatially arrayed material objects.

82. Another way of making the same sort of point is to note that although sensible qualities have location in the ways we have been investigating, it is the object and not the qualities which are emplaced, i.e., things but not qualities occupy space. One would not ask about the volume of the warmth in a warm object. Compare, in this respect, P. F. Strawson: "it is the things themselves, and not the processes they undergo, which are the primary *occupiers* of space, the possessors not only of spatial location, but of spatial *dimensions*." P. F. Strawson, *Individuals*, (London: Routledge, 1959), 57.

83. In *Individuals*, P. F. Strawson considers this as a sign of material objects: "We might regard it as a necessary condition of being a material body that it should tend to exhibit some felt resistance to touch." Strawson, *Individuals*, 39. But he gives up the condition, in the interest of full generality, as being too stringent. It is not clear to me that he need have given up this condition, nor that the generality he buys in doing so is at all useful to his analysis.

84. We might say the three aspects of objecthood of interest here are solidity, individuality, and particularity. The first two, I will claim, are known interactively (comportmentally) while the last is known informationally (sensibly). But, of course, knowing the particularity of the thing assumes knowledge of its individuality.

85. I am using this term in a more narrow sense than is usual to mean material particulars.

86. There is a kind of continuum of dependence here which we all tacitly recognize. Since everything is causally related, and every alteration has some degree of effect on surroundings, whether we count something as independent will depend on the scale of our interest. This should be obvious upon reflection: what counts as an independent object of interest, as an individual, for a molecular biologist is different from that which counts as an individual for a population biologist. My point here is that the scale of our interest can often be traced back to the active body, to a comportmental, behavioral interest in the world.

87. Thus stated the criterion is evidently too broad: the bottom can of peaches in a supermarket display is a thing independent of that display, yet to remove it from its surroundings is to cause the pyramid to collapse. Evidently, the can was not independent of the pyramid, nor the pyramid independent of the can. Thus, for instance, when the pyramid is the object of interest, we might define its boundaries, and thereby its individuality, by noting what must be moved to move the pyramid intact, i.e., as an independent object. Likewise, considered as part of the pyramid, the can is not independent; yet as a can it must be considered an independent thing. Is this independence contradicted by the effects of its removal? Let us consider the case more carefully: we can remove the can to the left, right or forward, and the character of this motion effects the resulting collapse not at all. For the pyramid is dependent on the can only to the extent that its removal exposes the pyramid to already existing forces (e.g., gravity), but it is independent in that the manipulation of the can as such (i.e., the actual character and extent of the movement of the can) does not (to that degree, in that manner) affect the pyramid. Consider instead a case where the cans are welded together: here grasping and moving the bottom can affects the pyramid in a way dependent on the actual character of the manipulation (its direction, speed, etc.) of the grasped can. It should be noted that the disanalogy which is here brought out in no way depends on the fact that in the latter case the components are joined by a physical bond: the example would work just as well had the pyramid been held together by a sufficiently strong magnetic force.

88. It is worth noting that manipulation is only one of a whole range of practical doings, insertions of the body in the causal order, by which I will claim we come to know objects and their boundaries. Nor are all manipulations literally so—breathing is a form of manipulation of air which plays an important role in our experience and knowledge of an atmosphere. Further, an interaction need not be a manipulation to reveal solid boundaries, and in the case of, say, a mountain manipulation would not in any case be possible. Still, the (vague) edges or beginnings of a mountain might well be known in terms of the different techniques of motion which would be required to negotiate such solid ground.

89. That is, our access is not well described merely as witnessing some "part" of the sensory field moving with respect to some other "part": this would entail prior recognition of these parts as independent, and their relative motion would be at best, a confirmation and not a discovery of that independence. We would still be left with the question of the principle of the division into parts of the sensory field.

90. Can we imagine the being in whom the behavioral and the representational failed to match? Were he an agent at all, he certainly would not be a product of natural selection, nor, as I argue in section 4.2, could such disembodied representations be the product of our normal developmental process of assigning significance to sensory/visual experiences primarily in terms of their position in our living practices.

91. That this recognition depends on an ability both to differentiate one's body from material solids, and to distinguish between actions and events, is readily admitted. I do not plan to address that capacity in this work, as I have no reason to believe that my arguments here are significantly distorted by the step I have taken to accept knowledge of one's actions, and of the body and its bounds as a given. This even though I realize that such knowledge likely comes in the course of our practical involvement with the world (although very early in our development).

92. This is a difficult point, for I am arguing that what we *count* as the whole is just that which so reacts. But the uniformity of the cube is in fact only a simple version of the coherence of "object-parts" which grounds our recognition of an individual as such. A length of rope would serve as a different level of complication: moving an end of a rope does not move the rest of it uniformly except in the case where it is being pulled along at full length—but although uniformity of motion is not evident here, still the character of the manipulation would reveal itself only in that which we would count as part of (or importantly connected to) the rope. At the other end of the spectrum on which the cube and rope lie would be, say, a particularly cohesive patch of molasses. Manipulating this would only ground recognition of it as constituting a particular to the extent that it was sufficiently cohesive to react as a whole (albeit not uniformly, due to the distortions to which it might be subject). Questions of how we would nevertheless recognize it as a patch of molasses (a volume of the *same stuff*) will be left to section 4.6.

93. We see here the difficulty in treating individuals apart from the class of which they are members; what identifies a thing as an individual is somewhat relative to its sortal type. Our access to the criteria of identity for types will be discussed in section 4.6.

94. See, in this regard, Kurt Lewin and J. J. Gibson. Lewin's notion of "invitation-character" (*Aufforderungscharakter*) and Gibson's notion of "affordances" is very close to what I am urging. (Thanks to Jack Saunders for this reference).

95. Thus do *trompe l'oeil* fool. Perceptual clues do not define identity.

96. Ludwig Wittgenstein, *Philosophical Investigations*, translated by G. E. M. Anscombe (New York: MacMillan, 1958), §288. Against the notion that sensations cannot have qualities which make them the sensations they are, Wittgenstein writes: "'The smell is marvelous!' Is there any doubt whether it is the smell that is marvelous? Is it a property of the smell?—Why not?" Ludwig Wittgenstein, *Zettel*, translated by G. E. M. Anscombe (Berkeley: University of California Press, 1967), §551. Sensations have properties granted by their place within a language game.

97. Wittgenstein, *Investigations*, §270.

98. That is, as Davidson saw was necessary, there are no epistemic mediators between our awareness and the world, neither "given" nor "contentful experience;" instead our experience is always already conceptual—and these concepts are *themselves* directly epistemically open to the world in comportment.

99. Wittgenstein, *Investigations*, §270.

100. A slightly more detailed account of perceptual significance can be found in the Wittgenstein section of my "Wittgenstein and Rousseau on the Context of Justification."

101. Oliver Sacks, in his newest book *An Anthropologist on Mars* relates the case of a congenitally blind man who, at the age of 45 has his sight restored. Rather than actually being able to see anything, the man reported a traumatic and totally incomprehensible blur of experiences, which only began to (very slowly) make sense as he began to relate the experiences to the already understood categories of his sightless world. Although the nature of his perceptual signals did not change, the appearance of the world gained significance only through being placed in his life-world.

102. As important as it is to treat separately the *epistemic* import of these two sensitivities, tactile and kinaesthetic, their deliverances cannot generally be separated phenomenologically in experience. Sensations of temperature are perhaps closest to the "tactile" pole, and those of weight closest to the kinaesthetic, but textures all involve some combination of tactile sensitivity and motion.

103. We must be careful not to associate improving perceptions of the world with more sophisticated causal antecedents to that perception, as is the case when we don a pair of eyeglasses, or see through a microscope. For such "causes" must still be taken up into the conceptual before they qualify as perceptual experience at all. As Hacking makes clear, we must learn to see through a microscope: accuracy here, just as in all perception, depends upon the appropriateness of the interpretation and not merely on the quality of the impinging causes. More to the point for my purposes is Hacking's own insistence that learning to see must involve bodily activity—only physical interference with the imaged object can teach us what it signifies.

104. This could be put into Heidegger-ese as follows: the kinaesthetic-body is the locus of Dasein's openness to Being; or perhaps: activity is the form of Dasein's openness to Being.

105. As we saw as early as chapter 1, there are various ways in which experience could be deemed autonomous in this sense. The most obvious is to posit an arbitrarily determined set of concepts which are solely responsible for synthesizing sensory inputs into experience. (It is a version of this thesis which Davidson and his followers deny.)

106. Note that it won't do to point out that my argument reduces to "If this theory of experience is true, then there is intentionality" as a way of reinstating global skepticism á la Stroud (since I cannot know if it *is* true). A skeptical attitude of sufficient corrosiveness to threaten radical doubt must contain the posit which my theory denies, and thus an argument along these lines *would* be tendentious and circular. (This is to say, I am only

theoretically, *ex hypothesi*, unable to know the truth of my proposal regarding the architecture of our experience if we accept the *skeptic's* version of that architecture, which prevents me from *knowing* anything [in the relevant sense].) Of course, as I mentioned in the first chapter, for this same reason I cannot claim to have refuted skepticism. I have only denied that we must accept the theory of cognitive psychology which lends itself so easily to skeptical objections. And, indeed, my theory may *not* be true, but I don't think the matter is theoretically unsettlable.

107. I mean to refer here to the persistence of inside/outside metaphors in describing our epistemic relation to the world: *I* am *inside*, and the world is *outside*. On this metaphor there is an interface between me (my experienced self, the self of my experience) and the world, but it is an interface which by its structure insists on the autonomy of content, on the under-determination of "interior content" by exterior reality. (Often this is motivated by valid concerns about human freedom.) In this sense the epistemic subject does not dwell in the world, but outside of it, however well connected the two may be.

108. I am not, of course, the only one. See, for instance, Michael Williams' *Unnatural Doubts*.

109. There are interesting thoughts to be thought about the ways in which this claim may restrict our intentional connection to historical particulars, and whether the restrictions it may impose are appropriate to the phenomenon of historical thinking.

110. Although the issues treated in the present work owe their impetus to Heidegger's example, the account, while Heideggerian, is not exactly Heidegger's, and for various reasons I am explicitly *not* trying to represent Heidegger's views, nor argue on behalf of Heidegger. Partly this has to do with the intended audience of this work (for which reason I will not be quoting Heidegger much), and partly it has to do with my sense that in *Being and Time* Heidegger, in making Dasein the horizon of reality, ends up defending another species of internalism. Conversations with Richard Capobianco have led me to give Heidegger another chance, and so I am planning a work comparing later Heidegger (primarily "The Question Concerning Technology") with Emerson's "Nature," where I treat both as grappling in various similar, occasionally successful ways with Kantian Idealism.

111. Charles Guignon, *Heidegger and the Problem of Knowledge* (Indianapolis: Hackett, 1983), 25.

112. Martin Heidegger, *Being and Time*, translated by Macquarrie and Robinson (New York: Harper & Row, 1971), 95.

113. The myth of the river Lethe, as recounted by Plato, is one instance of such a notion. Our birth as embodied beings was preceded by our existence as thinkers, made forgetful by our transition to the obfuscatory physical world. Whatever remembering we can do in this life, whatever clarity of thought we can achieve, will be purchased by overcoming (not cooperating with) our bodies.

114. Maxine Sheets-Johnstone, *The Roots of Thinking* (Philadelphia: Temple University Press, 1990). The many implications of this insistence that thinking is fundamentally intertwined with embodiment are explored in, e.g., Maxine Sheets-Johnstone, ed., *Giving the Body Its Due*; Michael O'Donovan-Anderson, ed., *The Incorporated Self*; and Julia MacCannell, ed., *Thinking Bodies*.

115. Guignon, *Heidegger and the Problem of Knowledge*, 61, 39.

116. I attribute this phenomenon to a local spatio-temporal distortion (no doubt predicted somewhere in Einstein's theories) which ensures that Boston geography is radically non-euclidean. In fact, I have it on reliable authority that Boston is the only place that mathematics Ph.D.'s *can* map (although oddly enough, they are on average five times more likely to get lost than non-mathematicians).

117. This might be attributable to the fact that, while our cognitive maps were strictly speaking inaccurate, they were nevertheless functionally adequate. This would preserve the possibility of making the claim that our practical activity is always cognitively driven, where cognition is understood as symbolic representation and mapping. I suspect that counter examples could be found among the drivers of Boston, but even if not: surely the reliance on functional mapping takes us a significant distance toward recognizing the importance of embodied activity to cognition. But both the empirical and theoretical aspects of this debate so posed carry us away from the primary arguments of this work.

One further note: I certainly do not want to be seen as making the claim that our behavior/activity is not mindful. What I am questioning is whether "mindful" is best understood as "driven by cognitive maps."

118. Mark Okrent, *Heidegger's Pragmatism* (Ithaca, NY: Cornell University Press, 1988), 130.

119. Guignon, *Heidegger and the Problem of Knowledge*, 100.

120. Guignon, *Heidegger and the Problem of Knowledge*, 147.

121. Of course, in the context of these interventions, one becomes familiar, too, with the characteristic "look" or "feel" of an object with a given function; but it does not follow from this that being a hammer can as easily be defined by stereotypical sensual-descriptive qualities as by functional place. A hammer belongs in that sort if and only if it fits into the given functional niche; not everything which shares the "family resemblance" or "look" belongs to the proper functional niche. Such an object might be a counterfeit representation of a hammer, but not a real hammer.

122. Okrent, *Heidegger's Pragmatism*, 130.

123. Note that this is a global constraint on a realist account of kinds, and does not apply to every case of sorting. Language carries sortal and other concepts which can thus be learned without actual manipulation of or interference with an instance of that sort. But if it is true that such sorts are often defined in terms of functional place or causal tendencies, then knowing the sort would imply knowing how to use or manipulate it. Further, it does seem that the arguments of the work so far indicate that (1) without interference and manipulation we could have made no nonarbitrary divisions in the sensory field, and (2) at least some sorts are *defined* in terms of the nature of the interventions used to sort them. Without intervention one would not have access to the matrix in terms of which these sorts were defined.

124. Okrent, *Heidegger's Pragmatism*, 149.

125. This is absolutely *not* to imply that thought or consciousness is somehow built up from mindless behavior. I am not offering an explanation of the origins of mindedness or thinking itself. If I *were* to do so it would not be in terms of mindless parts somehow constituting a minded whole, for this would be another instance of the "social contract mistake" of trying to make a (social) whole from (non-social) parts, or the "qualia mistake" of trying to get mental content from physical cause. (See chapter 3, above, and my "Wittgenstein and Rousseau on the Context of Justification.")

I *do* claim to be uncovering some conditions for contentful thinking by laying forth the practical origins of the empirical concepts which structure language and knowledge and belief. This is to say, in the context of what comes immediately above, that while language can indicate a cognitive, contemplative relation to the world, this is not a relation which supplants, transcends, or which is somehow independent of embodied access. In general, those philosophies which try to analyze mind in terms of body (i.e., behaviorism) or body in terms of mind (i.e., idealism) in point of fact just jettison mind and body in turn. The real trick is to be anti-dualist *and* anti-reductionist.

126. On these issues see, e.g., David Wiggins, *Sameness and Substance*, T. E. Wilkerson, *Natural Kinds* and "Species, Essences and the Names of Natural Kinds," John Dupre, "Wilkerson on Natural Kinds," Crawford Elder, "Realism, Naturalism and Culturally Generated Kinds," Herbert Granger, "Aristotle's Natural Kinds," and Chenyang Li, "Natural Kinds: Direct Reference, Realism and the Impossibility of Necessary *a posteriori* Truth."

127. At least, if it is not to be a bare metaphysical assertion with no epistemic or scientific value, it must be deeply connected to the causal, behavioral, functional propensities of members of the sort.

128. Don Ihde, *Technology and the Lifeworld*, (Bloomington, IN: Indiana University Press, 1990), 34.

129. This sort of critique might be more convincing against the theory-laden perception model of epistemology, as we could only see the world through the distorting lenses of our own interests. On the epistemic model I am proposing we can admit that our interests will have an effect on what we notice; our perspective may open us to certain aspects of the world, and dispose us to ignore other features, but in any case does not irretrievably cover over the world nor structure the world to fit the perspective we adopt.

130. E. O. Wilson's fumigation of isolated cypress trees to study the pattern by which species populate islands is one instance of intentional manipulation in the service of knowledge, our dumping of vast amounts of "greenhouse gasses" into the atmosphere may be an instance of unintentional manipulation which may yet result in great deal of knowledge.

131. David Resnik, "Hacking's Experimental Realism," *Canadian Journal of Philosophy* 24, no. 3 (September 1994): 400.

132. Ian Hacking, *Representing and Intervening* (Cambridge: Cambridge University Press, 1983), 173.

133. Hacking, *Representing and Intervening*, 262.

134. And recall that in the Heideggerian context, knowing need not mean "knowing that," but may also take the form of coping with. Thus to know the causal tendencies of a thing may mean to be able to interact usefully and purposefully with it.

135. Hacking, *Representing and Intervening*, 262.

136. Hacking, *Representing and Intervening*, 272.

137. Both charges are to be found in "Hacking's Experimental Realism" by David Resnik and "Hacking's Experimental Realism: An Untenable Middle Ground" by Richard Reiner and Robert Pierson.

138. Richard Reiner and Robert Pierson, "Hacking's Experimental Realism: An Untenable Middle Ground," *Philosophy of Science* 62 (1995): 67.

139. David Resnik makes the same move by distinguishing between two different senses of "use": "We can distinguish between two additional senses of the word 'use': an epistemic sense and a non-epistemic one. . . . When we say that we 'use an equation to

calculate probabilities' we are using 'use' in the epistemic sense; when we say that we 'use oxygen in respiration' we are using 'use' in the non-epistemic sense." Resnik, "Hacking's Experimental Realism," 405-6. Bodily interaction is a prime example of a non-epistemic sense of "use." Resnik argues that experimentation is in part an epistemic activity, and as such, must rely on theories and the justification of theories. "We cannot claim to use a theoretical entity as a tool for inquiry unless we also claim that it explains our experimental successes. But how do we explain those successes? By appealing to theories that describe the causal processes and structures used in our experiments." Resnik, "Hacking's Experimental Realism," 409.

140. Hacking's unclarity and imprecision at such argumentative junctures opens him to the (I still think unfair) criticisms of the writers already mentioned. I cannot be sure, for this reason, that placing Hacking in the context of my own work is true to *Representing and Intervening* as it stands. I do think it is consonant with his overall intentions for that book.

141. Resnik, "Hacking's Experimental Realism," 396.

142. It "saves" the phenomena (allows that intervention is a part of knowledge gathering) but analyzes its place and role in knowledge gathering in a way which invites anti-realism. For according to the "orthodox" theories of scientific method, our interpretation of the phenomena produced by intervention must be guided by theories, concepts, etc., and insofar as the interpreting theories/concepts cannot themselves be attributed to the world (as I argued at great length in chapter 3), neither can the knowledge which results from the interpretive process.

143. Nor, I think, do Hacking's analyses, if properly understood.

144. Hacking, *Representing and Intervening*, 189.

145. Hacking, *Representing and Intervening*, 204-5.

146. Of course, it is no criticism of this model to adopt a bit of epistemic humility: we cannot know if or when we know *everything* about electrons, in either of the senses of "know." We may not be distinguishing between balpeen-electrons and gavel-electrons, because we have not yet encountered them in the causal circumstances in which their functional-interactive differences become apparent. We can acknowledge our epistemic limitations, our finite grasp of the world, without despairing of the reliability and accuracy of our current know-how (and its requisite know-that). And insofar as interaction is a mode of epistemic *openness* to the world, the results of which can be attributed to the actual structure of the world, there is no reason to suppose that our continued interactions with the objects of the world, whether middle-sized or formerly hypothetical, will not lead us to a greater, more thorough and increasingly appropriate understanding of the world and its objects.

Conclusion

This work began, and has been implicitly concerned throughout, with the question of the possibility of truth. I put the issue this way because I do not intend in this work to actually provide an account of truth itself (other than to say, as I have already noted, that truth must be considered to be an assessment of the appropriateness of the relation of "aboutness" which our statements and thoughts bear to the world). Rather I will argue that the epistemological framework I have outlined meets the criteria for the possibility of a (realist) theory of truth, that is, prepares the ground for such an account to be given.

The general, and well-known, challenges facing any account of truth can be put succinctly: as a measure of the appropriateness of statements to their objects (subjects) truth loses its sense if we *define* assertions as appropriate by connecting language use too deeply with ontologization, by making linguistic meaning the agent of individuation. On the other side of the coin, truth becomes incomprehensible (even as a regulative ideal) if we imagine the "reality" by which assertions are to be measured to be humanly inaccessible (what Dummett calls "verification transcendent"). Reality and its significance must be available to human conception if any assessment of its relation to linguistic assertions is to be made. I argued in chapters 2 and 4 that no sensation based, single-mode theory of our epistemic access to the world can meet these two desiderata, for no account of our access to "appearances" can account for our knowledge of individuals without defining one in terms of the other (and therefore tending to the ideal); but to restrict our access to appearance is just to make the identities of natural (actual) individuals verification transcendent. Chapters 3 and 4 argued further that neither of these two options can provide a convincing account of the "aboutness" of language, for in the first case one would not be talking *about* some actual entity, but defining an ideal one, and in the second, although we might posit some occult power to account for language's (and thought's) bearing on reality, this is not a kind of "aboutness" that could ever play a role in the determinations of truth.

Any account of aboutness, of the possibility of reference for language and thought, must (1) provide an account not just of our capacity to gather (and interpret) perceptual information, but also of our ability to individuate and sort objects as such, (2) be able to deny that the phenomenal order of sensory knowledge is epistemically closed by providing the resources to attribute to the world the limitation and direction of our experience, and (3) provide some justification for the belief that our empirical knowledge is, or will tend to be, appropriate to the world and its structure.

The root of my thesis lies in the claim that our knowledge of the structured, physical world cannot be entirely accounted for in terms of the structured perceptual field, for our capacity to negotiate the world is related but not reducible to sensory/perceptual information. The perceptual psychology which accounts for this relation between information and activity is in certain respects identical to the usual (de-transcendentalized) neo-Kantian model: concepts are drawn passively into operation in the course of perceptual experience, and the particular nature of our awareness of the world is a result of the synthesis of the world's effect on us by our stock of empirical concepts. But I am claiming that these concepts are open to the world in virtue of, and their particular role in the synthetic operations of our perceptual mechanism is attributable to, the activity of the embodied self in the physical world. (see 4.3, 4.4) The embodied, active presence of the knowing self in the world doesn't just account for a level of access to the world, but also allows a level of confidence in the appropriateness of perceptual knowledge to the world, which sensation alone cannot provide.

Treating the body and bodily activity as a mode of epistemic access to the world allows for us to understand and account for (1) our epistemic access to the physical structure of reality (see 4.3), (2) our capacity to individuate particulars and determine sorts (see 4.3, 4.6), (3) the epistemic openness of the phenomenal order and the possibility of attributing the basic structure of that order (e.g., the divisions of the sensory field) to the actual structure of the world (see 4.3, 4.6), and (4) the likelihood that further revisions and modifications of the structure of the phenomenal order will be increasingly appropriate to the causal order (see 4.6).

The epistemic resources of the kinaesthetic-body allow the physical world to provide epistemic friction for empirical thinking, and *this* metaphor suggests the existence of a real congress between thought and world, overturning metaphors of cognitive confinement. We might say that our representational schema, our conceptual content, is coupled to the world by the transmission of comportment. Insofar as this is true, I claim to have provided epistemic grounds for a realistic theory of truth, not restricted to claiming only the maximal or ideal coherence of experience as its goal.

Bibliography

Adams, Henry. *The Education of Henry Adams*. New York: Modern Library, 1934.

Addis, Laird. "Intrinsic Reference and the New Theory." In *Midwest Studies in Philosophy XIV: Contemporary Perspectives in the Philosophy of Language*, edited by Peter A. French, Theodore E. Uehling and Howard K. Wettstein. Minneapolis, MN: University of Minnesota Press, 1990.

Adorno, Theodor, and Max Horkheimer. *Dialectic of Enlightenment*. Translated by John Cumming. New York: Continuum, 1989.

Akins, Kathleen. "Of Sensory Systems and the 'Aboutness' of Mental States." *Journal of Philosophy* 93, no. 7 (July 1996): 337-72.

Alcoff, Linda, and Elizabeth Potter, eds. *Feminist Epistemologies*. New York: Routledge, 1993.

Almeder, Robert. "Fallibilism and Ultimate Irreversible Opinion." *American Philosophical Quarterly* 9 (1975).

Anscombe, Elizabeth. *From Parmenides to Wittgenstein*. Minneapolis: University of Minnesota Press, 1981.

Appiah, Anthony. *Necessary Questions*. Englewood Cliffs, NJ: Prentice Hall, 1989.

———. *For Truth in Semantics*. Oxford: Basil Blackwell, 1986.

Aristotle. *Categories and De Interpretatione*. Translated by J. L. Ackrill. Oxford: Oxford University Press, 1963.

———. *Metaphysics*. Translated by Hippocrates G. Apostle. Grinnell, IA: Peripatetic Press, 1979.

———. *Nichomachean Ethics*. Translated by T. H. Irwin. Indianapolis: Hackett, 1985.

Armstrong, D. M. "Pereption and Belief." In *Perceptual Knowledge*, edited by Jonathan Dancy. Oxford: Oxford University Press, 1988.

Austin, J. L. "Performative Utterances." In *The Philosophy of Language*, edited by A. P. Martinich. Oxford: Oxford University Press, 1985.

Ayer, A. J. "Can There Be a Private Language?" In *The Philosophy of Language*, edited by A. P. Martinich. Oxford: Oxford University Press, 1985.

Ayers, Michael. *Locke*. London: Routledge, 1991.

———. "Substance: Prolegomena to a Realist Theory of Identity." *Journal of Philosophy* 9 (1991): 69-90.

Berger, Peter, and Thomas Luckmann. *The Social Construction of Reality.* New York: Anchor Doubleday, 1966.

Bilgrami, Akeel. *Belief and Meaning.* Cambridge, MA: Blackwell, 1992.

———. "Realism without Internalism: a Critique of Searle on Intentionality" *Journal of Philosophy* 86, no.2 (1989): 57-72.

Blackburn, Simon. "Manifesting Realism" In *Midwest Studies in Philosophy XIV: Contemporary Perspectives in the Philosophy of Language,* edited by Peter A. French, Theodore E. Uehling, and Howard K. Wettstein. Minneapolis, MN: University of Minnesota Press, 1990.

———. "Truth, Realism, and the Regulation of Theory" In *Midwest Studies in Philosophy V: Studies in Epistemology,* edited by Peter A. French, Theodore E. Uehling, and Howard K. Wettstein. Minneapolis, MN: University of Minnesota Press, 1981.

Blackburn, Thomas. "The Elusiveness of Reference" In *Midwest Studies in Philosophy XII: Realism and Antirealism,* edited by Peter A. French, Theodore E. Uehling, and Howard K. Wettstein. Minneapolis, MN: University of Minnesota Press, 1988.

Blanchette, Patricia. "Fregean Thoughts and Indexicals." *CSLI Report* no. 88-134 (November 1988).

Block, Ned, ed. *Readings in the Philosophy of Psychology,* vols. 1&2. Cambridge, MA: Harvard University Press, 1981.

Boyd, Richard. "The Current Status of Scientific Realism" In *Scientific Realism,* edited by Jarett Leplin. Berkeley: University of California Press, 1984.

Brandom, Robert. "Pragmatism, Phenomenalism and Truth Talk" In *Midwest Studies in Philosophy XII: Realism and Antirealism,* edited by Peter A. French, Theodore E. Uehling, and Howard K. Wettstein. Minneapolis, MN: University of Minnesota Press, 1988.

Braun, David. "Empty Names." *Noûs* XXXII, no.4 (December 1993).

Brown, Curtis. "Internal Realism: Transcendental Idealism" In *Midwest Studies in Philosophy XII: Realism and Antirealism,* edited by Peter A. French, Theodore E. Uehling, and Howard K. Wettstein. Minneapolis, MN: University of Minnesota Press, 1988.

Browning, Douglas. *Ontology and the Practical Arena.* College Park, PA: Penn State Press, 1990.

Butler, Judith. *Bodies That Matter.* New York: Routledge, 1993.

Casey, Edward. *Getting Back Into Place.* Indianapolis: Indiana University Press, 1993.

Cavell, Stanley. *The Claim of Reason.* Oxford: Oxford University Press, 1979.

———. *Must We Mean What We Say?* Cambridge: Cambridge University Press, 1976.

Cobb-Stevens, Richard. *Husserl and Analytic Philosophy.* Dordrecht: Kluwer Academic Publishers, 1990.

Dancy, Jonathan, ed. *Contemporary Epistemology.* Oxford: Basil Blackwell, 1985.

———. *Perceptual Knowledge.* Oxford: Oxford University Press, 1988.

Davidson, Donald. "A Coherence Theory of Truth and Knowledge" In *Truth and Interpretation,* edited by Ernest LePore. London: Basil Blackwell, 1986.

———. "On the Very Idea of a Conceptual Scheme" In *Inquiries into Truth and Interpretation.* Oxford: Oxford University Press, 1984.

———. *Inquiries into Truth and Interpretation.* Oxford: Oxford University Press, 1984.

Dennett, Daniel. "The Cartesian Theatre." Lecture at Yale University, 1992.

Descartes, Rene. *The Collected Writings of Descartes.* Translated by John Cottingham,

Robert Stoothoff, and Dugald Murdoch. Cambridge: Cambridge University Press, 1984.

Devitt, Michael. "Against Direct Reference." In *Midwest Studies in Philosophy XIV: Contemporary Perspectives in the Philosophy of Language*, edited by Peter A. French, Theodore E. Uehling, and Howard K. Wettstein. Minneapolis, MN: University of Minnesota Press, 1990.

———. *Realism and Truth*, 2d ed. Oxford: Basil Blackwell, 1991.

Diamond, Cora. *The Realistic Spirit*. Cambridge, MA: MIT Press, 1991.

Diggins, John Patrick. *The Promise of Pragmatism*. Chicago: University of Chicago Press, 1994.

Donellan, Keith. "Proper Names and Identifying Descriptions." In *Naming, Necessity and Natural Kinds*, edited by Stephen P. Schwartz. Ithaca: Cornell University Press, 1977.

———. "Reference and Definite Descriptions." In *The Philosophy of Language*, edited by A. P. Martinich. Oxford: Oxford University Press, 1985.

Dretske, Fred. *Explaining Behavior*. Cambridge, MA: MIT Press, 1988.

———. "Sensation and Perception." In *Perceptual Knowledge*, edited by Jonathan Dancy. Oxford: Oxford University Press, 1988.

Dreyfus, Hubert. *Being-in-the-World*. Cambridge, MA: MIT Press, 1991.

———. *What Computers Still Can't Do*. Cambridge, MA: MIT Press, 1993.

Dummett, Michael. *Frege: Philosophy of Language*, 2d ed. Cambridge, MA: Harvard Universiy Press, 1981.

———. *The Interpretation of Frege's Philosophy*. Cambridge, MA: Harvard University Press, 1981.

Dupre, John. "Wilkerson on Natural Kinds." *Philosophy* 64, (1989): 248-251.

Eagleton, Terry. *The Ideology of the Aesthetic*. Oxford: Basil Blackwell, 1990.

Elder, Crawford. "Realism, Naturalism and Culturally Generated Kinds." *The Philosophical Quarterly* 39, no. 157 (1989): 425-444.

Evans, Gareth. "The Causal Theory of Names." In *The Philosophy of Language*, edited by A. P. Martinich. Oxford: Oxford University Press, 1985.

———. *The Varieties of Reference*. Oxford: Oxford University Press, 1982.

Fine, Arthur. "The Natural Ontological Attitude." In *Scientific Realism*, edited by Jarett Leplin. Berkeley: University of California Press, 1984.

Flanagan, Owen. *The Science of The Mind*, 2d ed. Cambridge, MA: MIT Press, 1991.

Frege, Gottlob. *Begriffsschrift*. In *Frege and Godel: Two Fundamental Texts in Mathematical Logic*, edited by Jean van Heijenoort. Cambridge, MA: Harvard University Press, 1970.

———. *Philosophical and Mathematical Correspondence*. Chicago: University of Chicago Press, 1980.

———. *Posthumous Writings*. Chicago: University of Chicago Press, 1980. (Includes: "Function and Concept" and "The Thought".)

———. "On Sense and Meaning." In *The Philosophy of Language*, edited by A. P. Martinich. Oxford: Oxford University Press, 1985.

Friedman, Michael. *Kant and the Exact Sciences*. Cambridge: Harvard University Press, 1992.

Gadamer, Hans-Georg. *Truth and Method*, 2d ed. Translated by Weisenheimer and Marshall. New York: Crossroad, 1992.

Gavin, William. *William James and the Reinstatement of the Vague*. Philadelphia: Temple

University Press, 1992.

Gibson, J. J. *The Senses Considered as Perceptual Systems*. Boston: Houghton Mifflin, 1966.

Goldman, Alvin. "Discrimination and Perceptual Knowledge." In *Perceptual Knowledge*, edited by Jonathan Dancy. Oxford: Oxford University Press, 1988.

——, ed. *Readings in Philosophy and Cognitive Science*. Cambridge, MA: MIT Press, 1993.

Goodman, Nelson. *Ways of Worldmaking*. Indianapolis: Hackett, 1978.

Granger, Herbert. "Aristotle's Natural Kinds." *Philosophy* 64 (1989): 245-247.

Grice, H. P. "The Causal Theory of Perception." In *Perceptual Knowledge*, edited by Jonathan Dancy. Oxford: Oxford University Press, 1988.

——. *Studies in the Ways of Words*. Cambridge, MA: Harvard University Press, 1989.

——. "Utterer's Meaning and Intentions." In *The Philosophy of Language*, edited by A. P. Martinich. Oxford: Oxford University Press, 1985.

Guignon, Charles. *Heidegger and the Problem of Knowledge*. Indiannapolis: Hackett, 1983.

Hacking, Ian. "Experimentation and Scientific Realism" In *Scientific Realism*, edited by Jarrett Leplin. Berkeley: University of California Press, 1984.

——. *Representing and Intervening*. Cambridge: Cambridge University Press, 1983.

——. *Why Does Language Matter to Philosophy?* Cambridge: Cambridge University Press, 1975.

Hamlyn, D. W. *Schopenhauer: The Arguments of the Philosophers*. London: Routledge, 1980.

Hegel, G. W. F. *Elements of a Philosophy of Right*. Translated by H. B. Nisbet, edited by Allen Wood. Cambridge: Cambridge University Press, 1992.

——. *The Phenomenology of Spirit*. Translated by A.V. Miller. Oxford: Oxford Universiy Press, 1977.

Heidegger, Martin. *Being and Time*. Translated by John Macquarrie and Edward Robinson. New York: Harper & Row, 1962.

——. *Poetry, Language, Thought*. Translated by Albert Hofstadter. New York: Harper & Row, 1971.

——. "The Question Concerning Technology." In *Basic Writings*, edited by David Krell. New York: Harper & Row, 1977.

Henle, R. J. "Schopenhauer and Direct Realism." *Review of Metaphysics* 46 (September 1992): 125-140.

Hirsch, Eli. *Dividing Reality*. Oxford: Oxford University Press, 1993.

Humphrey, Ted B. "Schopenhauer and the Cartesian Tradition." *Journal of the History of Philosophy* 19 (April 1981): 191-212.

Ihde, Don. *Technology and the Lifeworld*. Bloomington, IN: Indiana University Press, 1990.

James, William. *Essays in Radical Empiricism*. New York: Longmans, Green, 1912.

——. *Pragmatism*. Indianapolis: Hackett, 1981.

——. *The Will to Believe*. New York: Dover, 1955.

——. *The Writings of William James*, edited by John J. McDermott. Chicago: University of Chicago Press, 1977.

Janaway, Christopher. *Schopenhauer*. Oxford: Oxford University Press, 1994.

Jay, Martin. *Downcast Eyes: The Denigration of Vision in 20th Century French Thought*. Berkeley: University of California Press, 1994.

Johnson, Mark. *The Body in the Mind*. Chicago: University of Chicago Press, 1987.

Kant, Immanuel. *Critique of Pure Reason*. Translated by Norman Kemp Smith. New York: St. Martin's, 1965.

———. *Opus Postumum*. Edited by Eckart Forster. Cambridge: Cambridge University Press, 1993.

———. *Prolegomena to Any Future Metaphysics*. Translated by James Ellington, Indianapolis: Hackett, 1972.

Kaplan, D. "Demonstratives." In *The Philosophy of Language*, edited by A. P. Martinich. Oxford: Oxford University Press, 1985.

Keil, Frank. *Concepts, Kinds and Cognitive Development*. Cambridge, MA: MIT Press, 1989.

Kornblith, Hilary. "Referring to Artifacts." *The Philosophical Review* LXXXIX, no. 1 (January, 1980): 109-14.

Kripke, Saul. *Naming and Necessity*. Cambridge, MA: Harvard University Press, 1980.

———. "On Rules and Private Language." In *The Philosophy of Language*, edited by A. P. Martinich. Oxford: Oxford University Press, 1985.

———. "Speaker's Reference and Semantic Reference." In *The Philosophy of Language*, edited by A. P. Martinich. Oxford: Oxford University Press, 1985.

Kuhn, Thomas. *The Structure of Scientific Revolutions*, 2d ed. Chicago: University of Chicago Press, 1970.

Lakoff, George. *Women, Fire and Dangerous Things: What Categories Reveal About the Mind*. Chicago: University of Chicago Press, 1987.

Laudan, Larry. "A Confutation of Convergent Realism." In *Scientific Realism*, edited by Jarrett Leplin. Berkeley: University of California Press, 1984.

Leder, Drew. *The Absent Body*. Chicago: University of Chicago Press, 1990.

Leplin, Jarrett, ed. *Scientific Realism*. Berkeley: University of California Press, 1984.

Levin, David Michael. *The Body's Recollection of Being*. London: Routledge, 1985.

———, ed. *Modernity and the Hegemony of Vision*. Berkeley: University of California Press, 1993.

Lewin, Kurt. *Principles of Topological Psychology*. Translated by Fritz Heider and Grace M. Heider. New York: McGraw Hill, 1969.

Li, Chenyang. "Natural Kinds: Direct Reference, Realism and the Impossibility of Necessary, *a posteriori* Truth." *Review of Metaphysics* 47 (December 1993): 261-276.

Lingis, Alphonso. *Foreign Bodies*. London: Routledge, 1994.

Locke, John. *An Essay Concerning Human Understanding*. New York: Dover, 1959.

Lycan, William, ed. *Mind and Cognition*. Cambridge, MA: Basil Blackwell, 1990.

MacCannell, Julia, ed. *Thinking Bodies*. Stanford: Stanford University Press, 1994.

McCumber, John. *The Company of Words: Hegel, Language and Systematic Philosophy*. Evanston, IL: Northwestern University Press, 1993.

MacDonald, G. F., ed. *Perception and Identity*. Ithaca, NY: Cornell Univerity Press, 1979.

McDowell, John. *Mind and World*. Cambridge, MA: Harvard University Press, 1994.

McGinn, Colin. *Mental Content*. Cambridge, MA: Basil Blackwell, 1989.

McMullin, Ernan. "A Case for Scientific Realism." In *Scientific Realism*, edited by Jarrett Leplin. Berkeley: University of California Press, 1984.

Margolis, Joseph. "A Biopsy of Recent Analytic Philosophy." *Philosophical Forum* 21, no. 3 (Spring 1995): 161-88.

Martinich, A. P., ed. *The Philosophy of Language*. Oxford: Oxford University Press, 1985.

Merleau-Ponty, Maurice. *The Phenomenology of Perception*. Translated by Colin Smith.
    London: Routledge, 1962.
Meyer, Leroy N. "Science, Reduction and Natural Kinds." *Philosophy* 64 (1989): 535-546.
Moore, John Alexander. *Science as a Way of Knowing: the Foundations of Modern
    Biology*. Cambridge, MA: Harvard University Press, 1993.
Mulhall, Stephen. *On Being in the World: Wittgenstein and Heidegger on Seeing Aspects*.
    London: Routledge, 1990.
Murphy, John P. *Pragmatism from Peirce to Davidson*. Boulder, CO: Westview, 1990.
Neale, Stephen. *Descriptions*. Cambridge, MA: MIT Press, 1990.
Neuhouser, Frederick. *Fichte's Theory of Subjectivity*. Cambridge: Cambridge University
    Press, 1990.
Nozick, Robert. "Knowledge and Skepticism." In *Perceptual Knowledge*, edited by
    Jonathan Dancy. Oxford: Oxford University Press, 1988.
Nussbaum, Martha. *Love's Knowledge*. Oxford: Oxford University Press, 1990.
O'Donovan-Anderson, Michael. "Certainty, Doubt and Truth: On the Nature, Scope and
    Degree of Doubt in Descartes' *Meditations*." *Lyceum* 5, no. 2 (Fall 1993): 19-65.
———, ed. *The Incorporated Self: Interdisciplinary Perspectives on Embodiment*. Lanham,
    MD: Rowman & Littlefield, 1996.
———. "Making Sense of Indexicals." *Lyceum* 4, no.1 (Spring 1992): 39-79.
———. "Wittgenstein and Rousseau on the Context of Justification." *Philosophy and Social
    Criticism* 22, no. 3 (May 1996): 75-92.
Okrent, Mark. *Heidegger's Pragmatism*. Ithaca, NY: Cornell University Press, 1988.
Peacocke, Christopher. "What Are Concepts?" In *Midwest Studies in Philosophy XIV:
    Contemporary Perspectives in the Philosophy of Language*, edited by Peter A. French,
    Theodore E. Uehling, and Howard K. Wettstein. Minneapolis, MN: University of
    Minnesta Press, 1990.
Peirce, Charles Sanders. *The Collected Papers of Charles Sanders Peirce*, Volumes 1-6,
    edited by Charles Hartshorne and Paul Weiss. Cambridge, MA: Harvard University
    Press, 1931-5; Volumes 7-8 edited by Arthur Burks, Cambridge, MA: Harvard
    University Press, 1958.
———. *The Philosophical Writings of Peirce*. Edited by Justus Buchler. New York: Dover,
    1955.
Prosch, Harry. *The Genesis of Twentieth Century Philosophy*. Garden City, NJ: Doubleday,
    1964.
Putnam, Hilary. "Explanation and Reference." In *Mind, Language, and Reality*.
    Cambridge: Cambridge University Press, 1975.
———. "Is Semantics Possible?" In *Mind, Language and Reality*. Cambridge: Cambridge
    University Press, 1975.
———. "The Meaning of 'Meaning.'" In *Mind, Language and Reality*. Cambridge:
    Cambridge University Press, 1975.
———. *Meaning and the Moral Sciences*. London: Routledge, 1978.
———. *Reason, Truth and History*. Cambridge: Cambridge University Press, 1981.
———. *Representation and Reality*. Cambridge, MA: MIT Press, 1988.
Quine, W. V. O. *Ontological Relativity*. New York: Columbia University Press, 1969.
———. "Two Dogmas of Empiricism." In *The Philosophy of Language*, edited by A. P.
    Martinich. Oxford: Oxford University Press, 1985.
———. *Word and Object*. Cambridge, MA: MIT Press, 1960.

Reiner, Robert and Robert Pierson. "Hacking's Experimental Realism: An Untenable Middle Ground." *Philosophy of Science* 62 (1995): 60-69.

Rescher, Nicholas. "Conceptual Idealism Revisited." *Review of Metaphysics* 44 (March 1991): 495-523.

Resnik, David. "Hacking's Experimental Realism." *Canadian Journal of Philosophy* 24, no.3 (September, 1994): 395-412.

Ricoeur, Paul. *Main Trends in Philosophy.* New York: Holmes & Meier, 1978.

Rorty, Richard. *The Consequences of Pragmatism.* Minneapolis: University of Minnesota Press, 1979.

——. *Essays on Heidegger and Others.* Cambridge: Cambridge Universiy Press, 1981.

——. *Objectivity, Relativism and Truth.* Cambridge: Cambridge University Press, 1991.

——. *Philosophy and the Mirror of Nature.* Princeton, NJ: Princeton University Press, 1979.

——. "Putnam and the Relativist Menace." *Journal of Philosophy* 90, no. 9 (September 1993): 443-461.

Rosen, Stanley. *The Limits of Analysis.* New Haven: Yale University Press, 1985.

Ruse, Michael. *Philosophy of Biology Today.* Albany: SUNY Press, 1988.

Russell, Bertrand. "Descriptions." In *The Philosophy of Language,* edited by A. P. Martinich. Oxford: Oxford University Press, 1985.

——. *Logic and Knowledge.* London: Unwin Hyman, 1956.

Sacks, Oliver. *An Anthropologist on Mars.* New York: Knopf, 1995.

Schwartz, Stephen P., ed. *Naming, Necessity and Natural Kinds.* Ithaca, NY: Cornell University Press, 1977.

Searle, John. "Proper Names." In *The Philosophy of Language,* edited by A. P. Martinich. Oxford: Oxford University Press, 1985.

Shell, Susan Meld. *The Embodiment of Reason: Kant on Spirit, Generation and Community.* Chicago: University of Chicago Press, 1996.

Siegfried, Charlene Haddock. *Chaos and Context: A Study in William James.* Athens, OH: Ohio University Press, 1985.

Sheets-Johnstone, Maxine. *The Roots of Thinking.* Philadelphia: Temple Universiy Press, 1990.

——, ed. *Giving the Body Its Due.* Albany, NY: SUNY Press, 1992.

Smith, John. *Purpose and Thought.* Chicago: Chicago University Press, 1978.

——. *Themes in American Philosophy.* New York: Harper & Row, 1970.

Sosa, Ernest. "Putnam's Pragmatic Realism." *Journal of Philosophy* 90, no. 12 (December, 1993): 605-626.

Strawson, P. F. *Individuals.* London: Routledge, 1959.

——. "Meaning and Truth." In *The Philosophy of Language,* edited by A. P. Martinich. Oxford: Oxford University Press, 1985.

——. "On Referring." In *The Philosophy of Language,* edited by A. P. Martinich. Oxford: Oxford University Press, 1985.

Suckiel, Ellen. *The Pragmatic Philosophy of William James.* Notre Dame, IN: University of Notre Dame Press, 1982.

Tarski, A. "The Semantic Conception of Truth." In *The Philosophy of Language,* edited by A. P. Martinich. Oxford: Oxford University Press, 1985.

Taylor, Charles. *Hegel.* Cambridge: Cambridge University Press, 1975.

——. "Sense Data Revisited." In *Perception and Identity,* edited by G. F. MacDonald.

Ithaca, NY: Cornell University Press, 1979.

———. "The Validity of Transcendental Arguments." *Proceedings of the Aristotelian Society* 79, (1978-9).

Thompson, Evan. *Colour Vision.* London: Routledge, 1995.

VanFraassen, Bas C. "To Save the Phenomena." In *Scientific Realism*, edited by Jarrett Leplin. Berkeley: University of California Press, 1984.

Wettstein, Howard. "Cognitive Significance Without Cognitive Content." *Mind* 97, no. 385 (January 1988).

Wilkerson, T. E. *Natural Kinds.* Avebury, VT: Avebury Press, 1995.

———. "Species, Essences and the Names of Natural Kinds." *Philosophical Quarterly* 43, no. 170 (1988): 1-19.

Williams, Bernard. *Descartes: The Project of Pure Enquiry.* New York: Penguin, 1978.

Williams, Michael. *Unnatural Doubts.* Princeton, NJ: Princeton University Press, 1996.

Wiggins, David. *Sameness and Substance.* Cambridge, MA: Harvard University Press, 1980.

Winch, Peter. *The Idea of a Social Science*, 2d ed. Atlantic Highlands, NJ: Humanities Press, 1990.

Wittgenstein, Ludwig. *Philosophical Investigations*, 3d ed. Translated by G. E. M. Anscombe. New York: MacMillan, 1958.

———. *Zettel.* Translated by G. E. M. Anscombe. Berkeley: University of California Press, 1967.

Wood, Allen. *Hegel's Ethical Thought.* Cambridge: Cambridge University Press, 1990.

Wright, Crispin. "Realism, Antirealism, Irrealism, Quasirealism." In *Midwest Studies in Philosophy XII: Realism and Antirealism*.edited by Peter A. French, Theodore E. Uehling, and Howard K. Wettstein. Minneapolis, MN: University of Minnesota Press, 1988.

Index

About the Author

Michael O'Donovan-Anderson holds a Ph.D. in philosophy from Yale University, and is a tutor at St. John's College, Annapolis. He is the editor of *The Incorporated Self: Interdisciplinary Perspectives on Embodiment* and author of several articles in philosophy and related disciplines.

DATE DUE

WITHDRAWN

GAYLORD

PRINTED IN U.S.A.

BD 201 .O36 1997

O'Donovan-Anderson, Michael.

Content and comportment